SYSTEM DESIGN INTERVIEW

Alex Xu

SYSTEM DESIGN INTERVIEW - AN INSIDER'S GUIDE

Copyright © 2020 Byte Code LLC

About the author:

Alex Xu is an experienced software engineer and entrepreneur. Previous-ly, he worked at Twitter, Apple, Zynga and Oracle. He received his M.S. from Carnegie Mellon University. He has a passion for designing and implementing complex systems.

For more information, contact hi@bytebytego.com

Editor: Paul Solomon

System Design Newsletter

Subscribe to the ByteByteGo weekly newsletter to get a Free System Design PDF (158 pages): blog.bytebytego.com

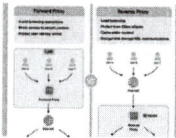

EP26: Proxy vs reverse proxy

In this issue, we will cover: Why is Nginx called a "reverse" proxy? CAP theorem How Does Live Streaming Platform Work? CDN Postman the API platform for...

ALEX XU   OCT 1   ♡ 225   💬 6   ↪

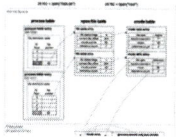

EP17: Design patterns cheat sheet. Also...

For this week's newsletter, we will cover: Design patterns cheat sheet 6 ways to turn code into beautiful architecture diagrams What is a File...

ALEX XU   JUL 30   ♡ 166   💬 7   ↪

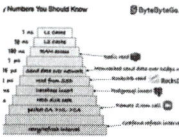

EP22: Latency numbers you should know. Also...

In this newsletter, we'll cover the following topics: Latency numbers you should know Microservice architecture Handling hotspot accounts E-commerce...

ALEX XU   SEP 3   ♡ 153   💬 9   ↪

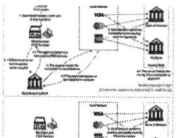

EP15: What happens when you swipe a credit card? Also...

For this week's newsletter, we will cover: How does VISA work when we swipe a credit card at a merchant's shop? What are the differences between bare...

ALEX XU   JUL 16   ♡ 141   💬 8   ↪

EP14: Algorithms you should know for System Design. Also...

In this newsletter, we'll cover the following topics: Algorithms you should know before taking System Design Interviews How to store passwords safely in...

ALEX XU   JUL 9   ♡ 185   💬 2   ↪

CONTENTS

FOREWORD

We are delighted that you have decided to join us in learning the system design interviews. System design interview questions are the most difficult to tackle among all the technical interviews. The questions require the interviewees to design an architecture for a software system, which could be a news feed, Google search, chat system, etc. These questions are intimidating, and there is no certain pattern to follow. The questions are usually very big scoped and vague. The processes are open-ended and unclear without a standard or correct answer.

Companies widely adopt system design interviews because the communication and problem-solving skills tested in these interviews are similar to those required by a software engineer's daily work. An interviewee is evaluated based on how she analyzes a vague problem and how she solves the problem step by step. The abilities tested also involve how she explains the idea, discusses with others, and evaluates and optimizes the system. In English, using "she" flows better than "he or she" or jumping between the two. To make reading easier, we use the feminine pronoun throughout this book. No disrespect is intended for male engineers.

The system design questions are open-ended. Just like in the real world, there are many differences and variations in the system. The desired outcome is to come up with an architecture to achieve system design goals. The discussions could go in different ways depending on the interviewer. Some interviewers may choose high-level architecture to cover all aspects; whereas some might choose one or more areas to focus on. Typically, system requirements, constraints and bottlenecks should be well understood to shape the direction of both the interviewer and interviewee.

The objective of this book is to provide a reliable strategy to approach the system design questions. The right strategy and knowledge are vital to the success of an interview.

This book provides solid knowledge in building a scalable system. The more knowledge gained from reading this book, the better you are equipped in solving the system design questions.

This book also provides a step by step framework on how to tackle a system design question. It provides many examples to illustrate the systematic approach with detailed steps that you can follow. With constant practice, you will be well-equipped to tackle system design interview questions.

SCALE FROM ZERO TO MILLIONS OF USERS

Designing a system that supports millions of users is challenging, and it is a journey that requires continuous refinement and endless improvement. In this chapter, we build a system that supports a single user and gradually scale it up to serve millions of users. After reading this chapter, you will master a handful of techniques that will help you to crack the system design interview questions.

Single server setup

A journey of a thousand miles begins with a single step, and building a complex system is no different. To start with something simple, everything is running on a single server. Figure 1-1 shows the illustration of a single server setup where everything is running on one server: web app, database, cache, etc.

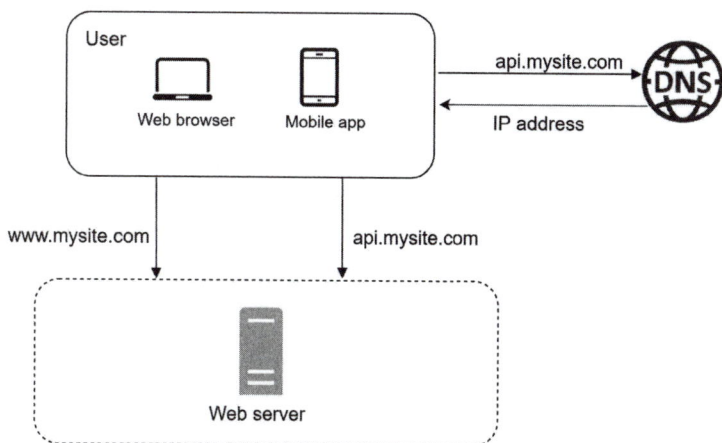

Figure 1-1

To understand this setup, it is helpful to investigate the request flow and traffic source. Let us first look at the request flow (Figure 1-2).

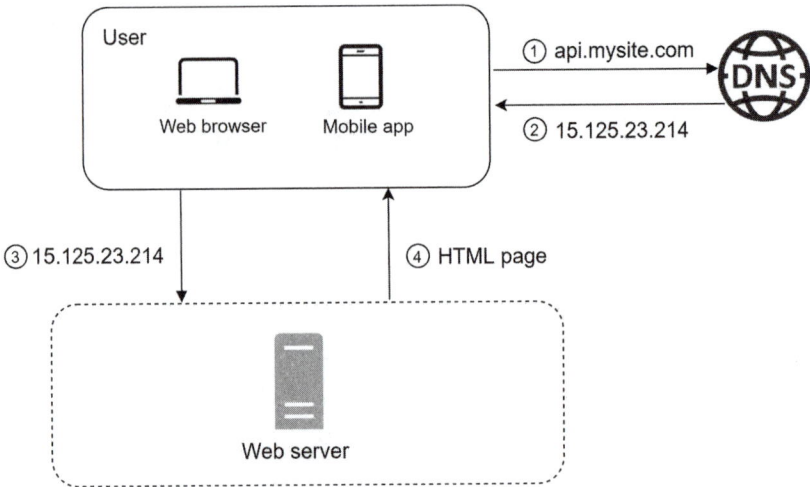

Figure 1-2

1. Users access websites through domain names, such as api.mysite. com. Usually, the Domain Name System (DNS) is a paid service provided by 3rd parties and not hosted by our servers.

2. Internet Protocol (IP) address is returned to the browser or mobile app. In the example, IP address 15.125.23.214 is returned.

3. Once the IP address is obtained, Hypertext Transfer Protocol (HTTP) [1] requests are sent directly to your web server.

4. The web server returns HTML pages or JSON response for rendering.

Next, let us examine the traffic source. The traffic to your web server comes from two sources: web application and mobile application.

- Web application: it uses a combination of server-side languages (Java, Python, etc.) to handle business logic, storage, etc., and client-side languages (HTML and JavaScript) for presentation.

- Mobile application: HTTP protocol is the communication protocol between the mobile app and the web server. JavaScript Object Notation (JSON) is commonly used API response format to transfer data due to its simplicity. An example of the API response in JSON format is shown below:

GET /users/12 – Retrieve user object for id = 12

```
{
    "id": 12,
    "firstName": "John",
    "lastName": "Smith",
    "address": {
        "streetAddress": "21 2nd Street",
        "city": "New York",
        "state": "NY",
        "postalCode": 10021
    },
    "phoneNumbers": [
        "212 555-1234",
        "646 555-4567"
    ]
}
```

Database

With the growth of the user base, one server is not enough, and we need multiple servers: one for web/mobile traffic, the other for the database (Figure 1-3). Separating web/mobile traffic (web tier) and database (data tier) servers allows them to be scaled independently.

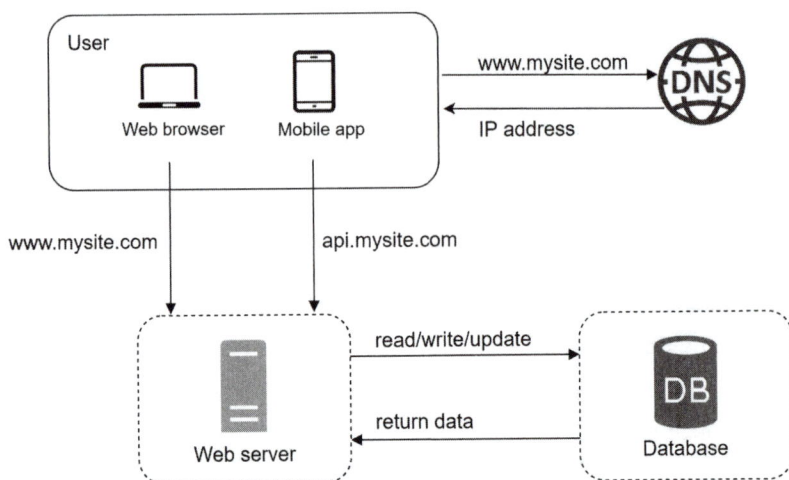

Figure 1-3

Which databases to use?

You can choose between a traditional relational database and a non-relational database. Let us examine their differences.

Relational databases are also called a relational database management system (RDBMS) or SQL database. The most popular ones are MySQL, Oracle database, PostgreSQL, etc. Relational databases represent and store data in tables and rows. You can perform join operations using SQL across different database tables.

Non-Relational databases are also called NoSQL databases. Popular ones are CouchDB, Neo4j, Cassandra, HBase, Amazon DynamoDB, etc. [2]. These databases are grouped into four categories: key-value stores, graph stores, column stores, and document stores. Join operations are generally not supported in non-relational databases.

For most developers, relational databases are the best option because they have been around for over 40 years and historically, they have worked well. However, if relational databases are not suitable for your specific use cases, it is critical to explore beyond relational databases. Non-relational databases might be the right choice if:

- Your application requires super-low latency.
- Your data are unstructured, or you do not have any relational data.
- You only need to serialize and deserialize data (JSON, XML, YAML, etc.).
- You need to store a massive amount of data.

Vertical scaling vs horizontal scaling

Vertical scaling, referred to as "scale up", means the process of adding more power (CPU, RAM, etc.) to your servers. Horizontal scaling, referred to as "scale-out", allows you to scale by adding more servers into your pool of resources.

When traffic is low, vertical scaling is a great option, and the simplicity of vertical scaling is its main advantage. Unfortunately, it comes with serious limitations.

- Vertical scaling has a hard limit. It is impossible to add unlimited CPU and memory to a single server.
- Vertical scaling does not have failover and redundancy. If one server goes down, the website/app goes down with it completely.

Horizontal scaling is more desirable for large scale applications due to the limitations of vertical scaling.

In the previous design, users are connected to the web server directly. Users will be unable to access the website if the web server is offline. In another scenario, if many users access the web server simultaneously and it reach-es the web server's load limit, users generally experience slower response or fail to connect to the server. A load balancer is the best technique to address these problems.

Load balancer

A load balancer evenly distributes incoming traffic among web servers that are defined in a load-balanced set. Figure 1-4 shows how a load balancer works.

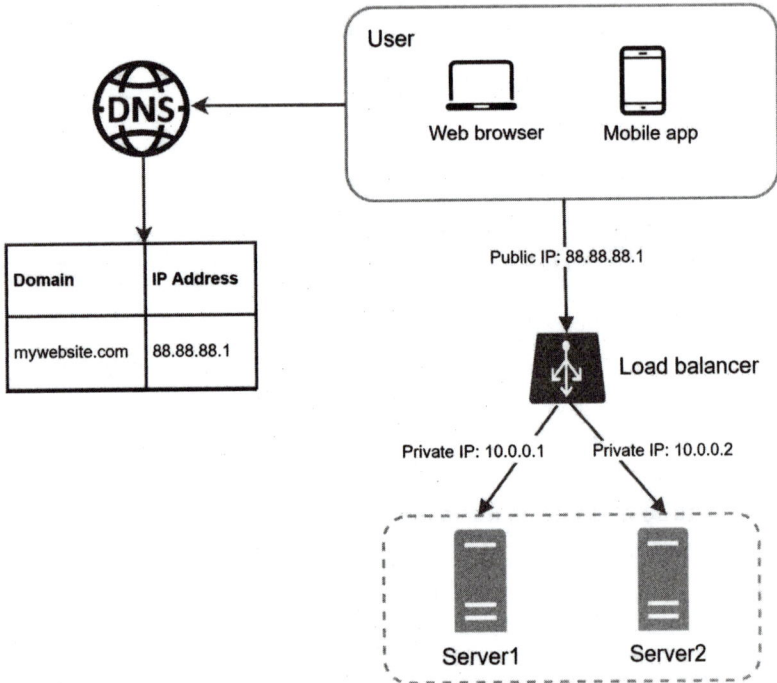

Figure 1-4

As shown in Figure 1-4, users connect to the public IP of the load balancer directly. With this setup, web servers are unreachable directly by clients anymore. For better security, private IPs are used for communication between servers. A private IP is an IP address reachable only between servers in the same network; however, it is unreachable over the internet. The load balancer communicates with web servers through private IPs.

In Figure 1-4, after a load balancer and a second web server are added, we successfully solved no failover issue and improved the availability of the web tier. Details are explained below:

- If server 1 goes offline, all the traffic will be routed to server 2. This prevents the website from going offline. We will also add a new healthy web server to the server pool to balance the load.

- If the website traffic grows rapidly, and two servers are not enough to handle the traffic, the load balancer can handle this problem gracefully. You only need to add more servers to the web server pool, and the load balancer automatically starts to send requests to them.

Now the web tier looks good, what about the data tier? The current design has one database, so it does not support failover and redundancy. Database replication is a common technique to address those problems. Let us take a look.

Database replication

Quoted from Wikipedia: "Database replication can be used in many database management systems, usually with a master/slave relationship between the original (master) and the copies (slaves)" [3].

A master database generally only supports write operations. A slave database gets copies of the data from the master database and only supports read operations. All the data-modifying commands like insert, delete, or update must be sent to the master database. Most applications require a much higher ratio of reads to writes; thus, the number of slave databases in a system is usually larger than the number of master databases. Figure 1-5 shows a master database with multiple slave databases.

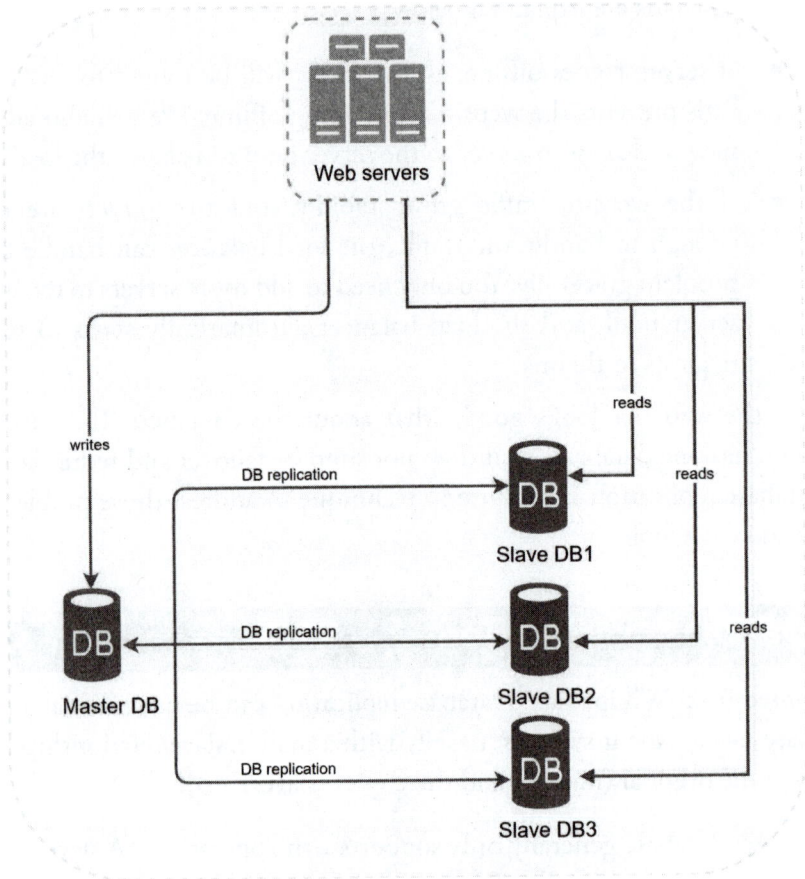

Figure 1-5

Advantages of database replication:

- Better performance: In the master-slave model, all writes and updates happen in master nodes; whereas, read operations are distributed across slave nodes. This model improves performance because it allows more queries to be processed in parallel.

- Reliability: If one of your database servers is destroyed by a natural disaster, such as a typhoon or an earthquake, data is still preserved. You do not need to worry about data loss because data is replicated across multiple locations.

- High availability: By replicating data across different locations, your website remains in operation even if a database is offline as you can access data stored in another database server.

In the previous section, we discussed how a load balancer helped to improve system availability. We ask the same question here: what if one of the databases goes offline? The architectural design discussed in Figure 1-5 can handle this case:

- If only one slave database is available and it goes offline, read operations will be directed to the master database temporarily. As soon as the issue is found, a new slave database will replace the old one. In case multiple slave databases are available, read operations are redirected to other healthy slave databases. A new database server will replace the old one.

- If the master database goes offline, a slave database will be promoted to be the new master. All the database operations will be temporarily executed on the new master database. A new slave database will replace the old one for data replication immediately. In production systems, promoting a new master is more complicated as the data in a slave database might not be up to date. The missing data needs to be updated by running data recovery scripts. Although some other replication methods like multi-masters and circular replication could help, those setups are more complicated; and their discussions are beyond the scope of this book. Interested readers should refer to the listed reference materials [4] [5].

Figure 1-6 shows the system design after adding the load balancer and database replication.

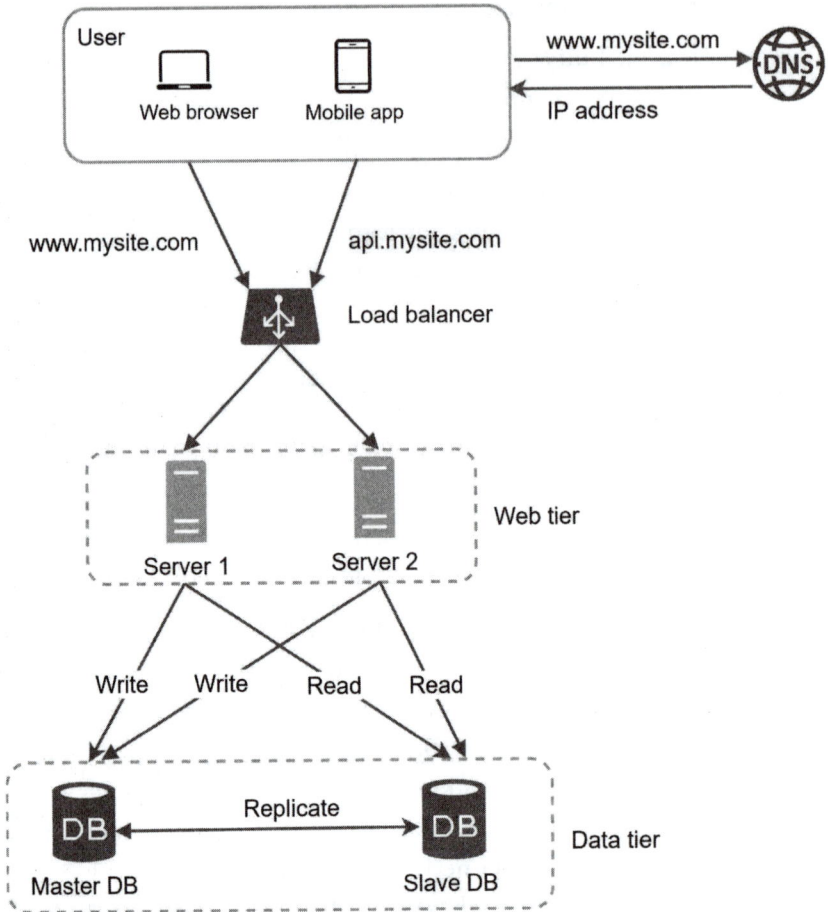

Figure 1-6

Let us take a look at the design:

- A user gets the IP address of the load balancer from DNS.

- A user connects the load balancer with this IP address.

- The HTTP request is routed to either Server 1 or Server 2.

- A web server reads user data from a slave database.

- A web server routes any data-modifying operations to the master database. This includes write, update, and delete operations.

Now, you have a solid understanding of the web and data tiers, it is time to improve the load/response time. This can be done by adding a cache layer and shifting static content (JavaScript/CSS/image/video files) to the content delivery network (CDN).

Cache

A cache is a temporary storage area that stores the result of expensive responses or frequently accessed data in memory so that subsequent requests are served more quickly. As illustrated in Figure 1-6, every time a new web page loads, one or more database calls are executed to fetch data. The application performance is greatly affected by calling the database repeatedly. The cache can mitigate this problem.

Cache tier

The cache tier is a temporary data store layer, much faster than the database. The benefits of having a separate cache tier include better system performance, ability to reduce database workloads, and the ability to scale the cache tier independently. Figure 1-7 shows a possible setup of a cache server:

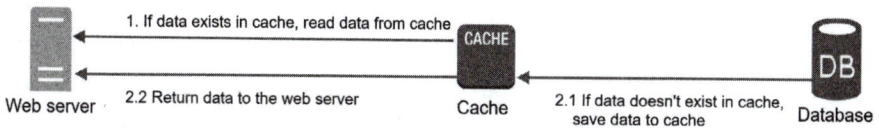

1. If data exists in cache, read data from cache

Web server   2.2 Return data to the web server   Cache   2.1 If data doesn't exist in cache,   Database
                                                                save data to cache

Figure 1-7

After receiving a request, a web server first checks if the cache has the available response. If it has, it sends data back to the client. If not, it queries the database, stores the response in cache, and sends it back to the client. This caching strategy is called a read-through cache. Other caching strategies are available depending on the data type, size, and access patterns. A previous study explains how different caching strategies work [6].

Interacting with cache servers is simple because most cache servers provide APIs for common programming languages. The following code snippet shows typical Memcached APIs:

```
SECONDS = 1
cache.set('myKey', 'hi there', 3600 * SECONDS)
cache.get('myKey')
```

Considerations for using cache

Here are a few considerations for using a cache system:

- Decide when to use cache. Consider using cache when data is read frequently but modified infrequently. Since cached data is stored in volatile memory, a cache server is not ideal for persisting data. For instance, if a cache server restarts, all the data in memory is lost. Thus, important data should be saved in persistent data stores.

- Expiration policy. It is a good practice to implement an expiration policy. Once cached data is expired, it is removed from the cache. When there is no expiration policy, cached data will be stored in the memory permanently. It is advisable not to make the expiration date too short as this will cause the system to reload data from the database too frequently. Meanwhile, it is advisable not to make the expiration date too long as the data can become stale.

- Consistency: This involves keeping the data store and the cache in sync. Inconsistency can happen because data-modifying operations on the data store and cache are not in a single transaction. When scaling across multiple regions, maintaining consistency between the data store and cache is challenging. For further details, refer to the paper titled "Scaling Memcache at Facebook" published by Facebook [7].

- Mitigating failures: A single cache server represents a potential single point of failure (SPOF), defined in Wikipedia as follows:

"A single point of failure (SPOF) is a part of a system that, if it fails, will stop the entire system from working" [8]. As a result, multiple cache servers across different data centers are recommended to avoid SPOF. Another recommended approach is to overprovision the required memory by certain percentages. This provides a buffer as the memory usage increases.

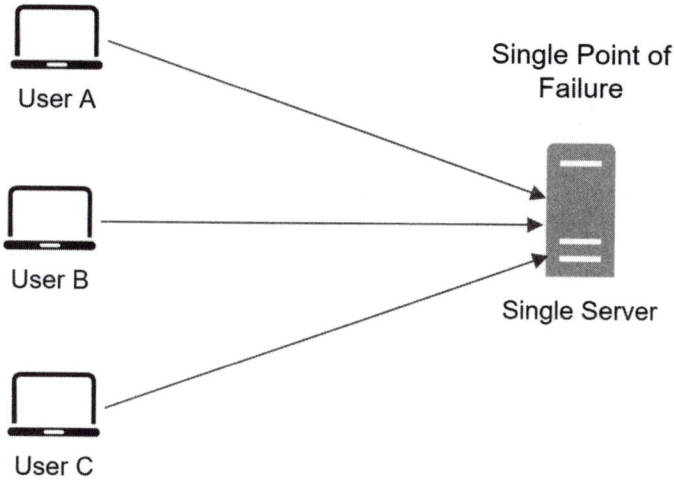

Figure 1-8

- Eviction Policy: Once the cache is full, any requests to add items to the cache might cause existing items to be removed. This is called cache eviction. Least-recently-used (LRU) is the most popular cache eviction policy. Other eviction policies, such as the Least Frequently Used (LFU) or First in First Out (FIFO), can be adopted to satisfy different use cases.

Content delivery network (CDN)

A CDN is a network of geographically dispersed servers used to deliver static content. CDN servers cache static content like images, videos, CSS, JavaScript files, etc.

Dynamic content caching is a relatively new concept and beyond the scope of this book. It enables the caching of HTML pages that are based on request path, query strings, cookies, and request headers. Refer to the article mentioned in reference material [9] for more about this. This book focuses on how to use CDN to cache static content.

Here is how CDN works at the high-level: when a user visits a website, a CDN server closest to the user will deliver static content. Intuitively, the further users are from CDN servers, the slower the website loads. For example, if CDN servers are in San Francisco, users in Los Angeles will get content faster than users in Europe. Figure 1-9 is a great example that shows how CDN improves load time.

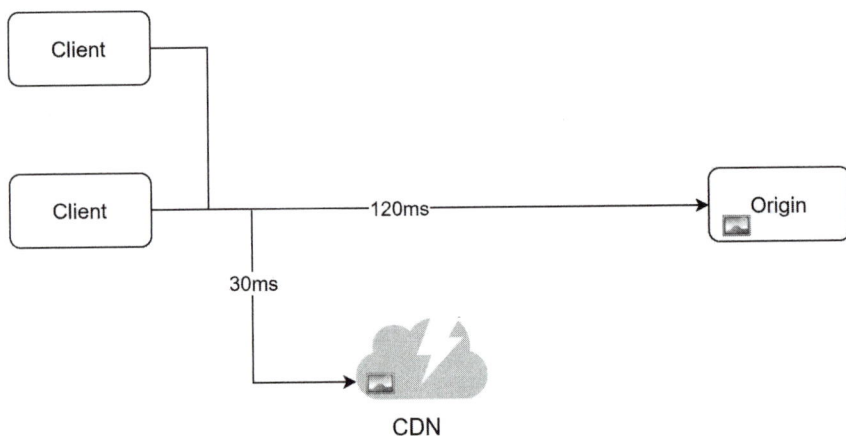

Figure 1-9

Figure 1-10 demonstrates the CDN workflow.

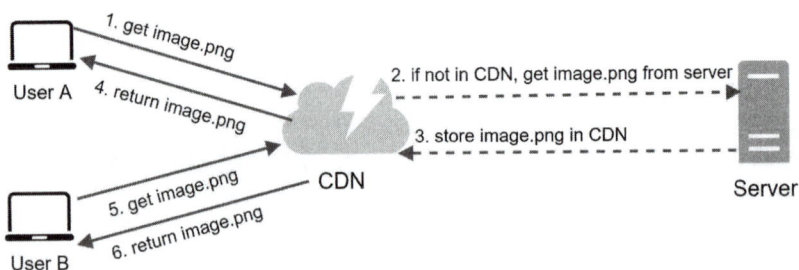

Figure 1-10

1. User A tries to get image.png by using an image URL. The URL's domain is provided by the CDN provider. The following two image URLs are samples used to demonstrate what image URLs look like on Amazon and Akamai CDNs:

   • https://mysite.cloudfront.net/logo.jpg

   • https://mysite.akamai.com/image-manager/img/logo.jpg

2. If the CDN server does not have image.png in the cache, the CDN server requests the file from the origin, which can be a web server or online storage like Amazon S3.

3. The origin returns image.png to the CDN server, which includes optional HTTP header Time-to-Live (TTL) which describes how long the image is cached.

4. The CDN caches the image and returns it to User A. The image remains cached in the CDN until the TTL expires.

5. User B sends a request to get the same image.

6. The image is returned from the cache as long as the TTL has not expired.

Considerations of using a CDN

• Cost: CDNs are run by third-party providers, and you are charged for data transfers in and out of the CDN. Caching infrequently used assets provides no significant benefits so you should consider moving them out of the CDN.

• Setting an appropriate cache expiry: For time-sensitive content, setting a cache expiry time is important. The cache expiry time should neither be too long nor too short. If it is too long, the content might no longer be fresh. If it is too short, it can cause repeat reloading of content from origin servers to the CDN.

• CDN fallback: You should consider how your website/application copes with CDN failure. If there is a temporary CDN outage, clients should be able to detect the problem and request resources from the origin.

- Invalidating files: You can remove a file from the CDN before it expires by performing one of the following operations:

  o Invalidate the CDN object using APIs provided by CDN vendors.

  o Use object versioning to serve a different version of the object. To version an object, you can add a parameter to the URL, such as a version number. For example, version number 2 is added to the query string: image.png?v=2.

Figure 1-11 shows the design after the CDN and cache are added.

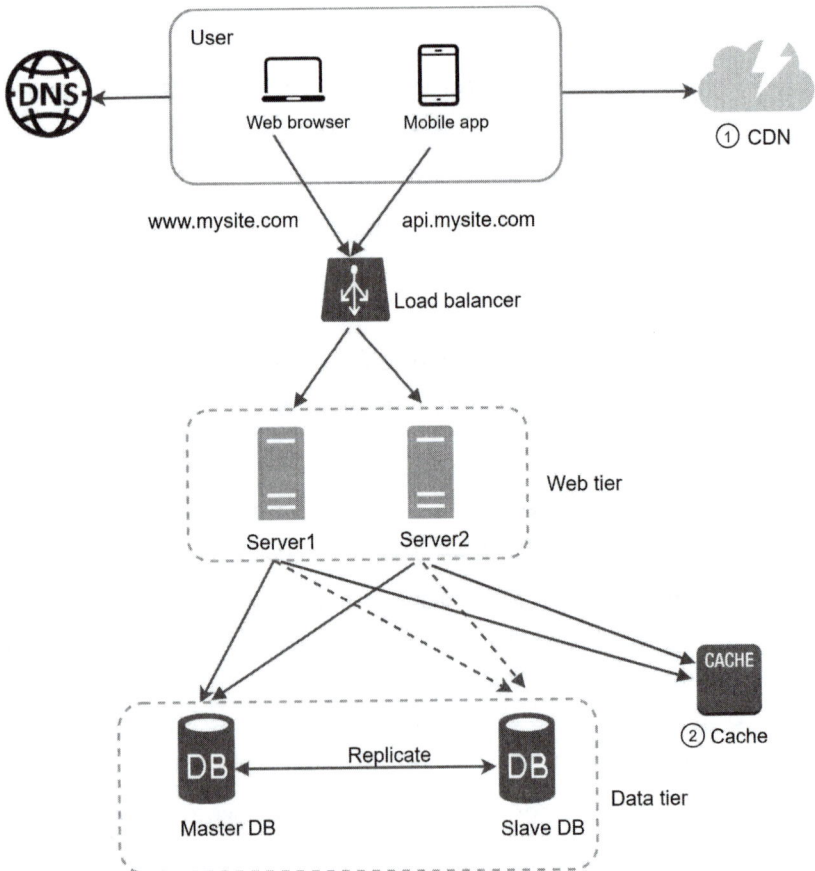

Figure 1-11

1. Static assets (JS, CSS, images, etc.,) are no longer served by web servers. They are fetched from the CDN for better performance.

2. The database load is lightened by caching data.

Stateless web tier

Now it is time to consider scaling the web tier horizontally. For this, we need to move state (for instance user session data) out of the web tier. A good practice is to store session data in the persistent storage such as relational database or NoSQL. Each web server in the cluster can access state data from databases. This is called stateless web tier.

Stateful architecture

A stateful server and stateless server has some key differences. A stateful server remembers client data (state) from one request to the next. A stateless server keeps no state information.

Figure 1-12 shows an example of a stateful architecture.

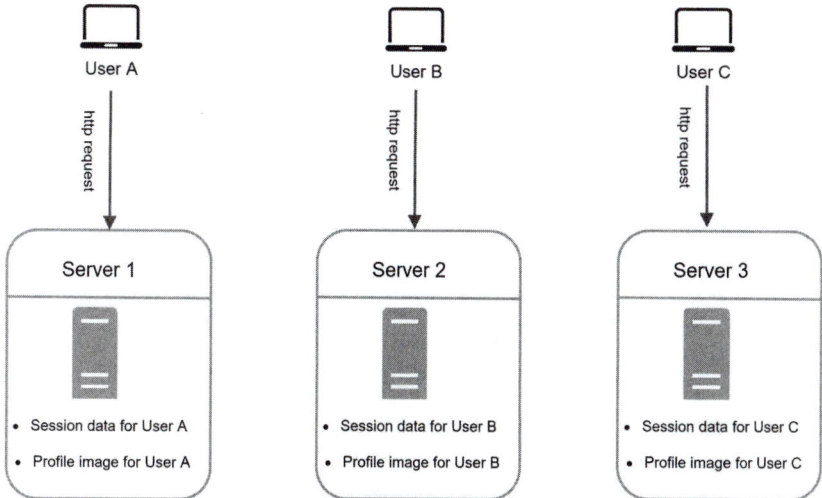

Figure 1-12

In Figure 1-12, user A's session data and profile image are stored in Server 1. To authenticate User A, HTTP requests must be routed to Server 1. If a request is sent to other servers like Server 2, authentication would fail because Server 2 does not contain User A's session data. Similarly, all HTTP requests from User B must be routed to Server 2; all requests from User C must be sent to Server 3.

The issue is that every request from the same client must be routed to the same server. This can be done with sticky sessions in most load balancers [10]; however, this adds the overhead. Adding or removing servers is much more difficult with this approach. It is also challenging to handle server failures.

Stateless architecture

Figure 1-13 shows the stateless architecture.

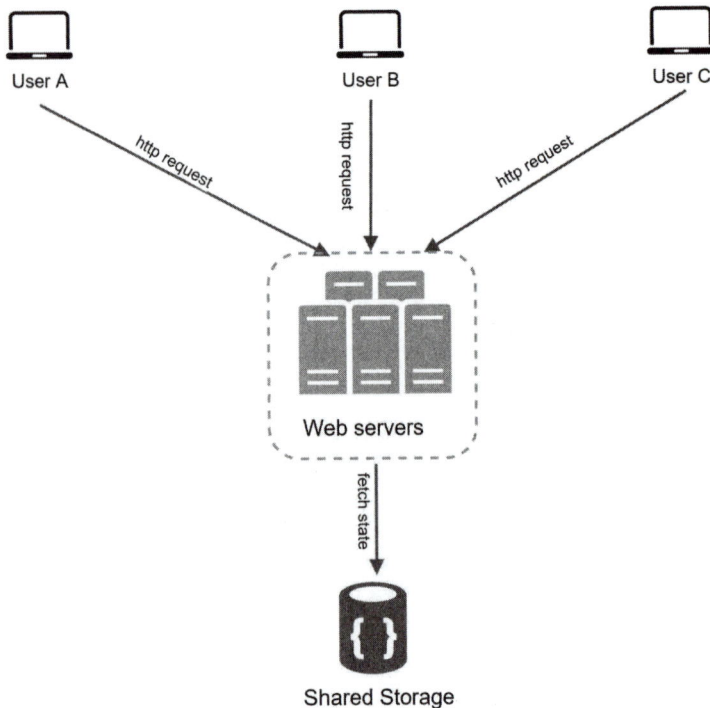

Figure 1-13

In this stateless architecture, HTTP requests from users can be sent to any web servers, which fetch state data from a shared data store. State data is stored in a shared data store and kept out of web servers. A stateless system is simpler, more robust, and scalable.

Figure 1-14 shows the updated design with a stateless web tier.

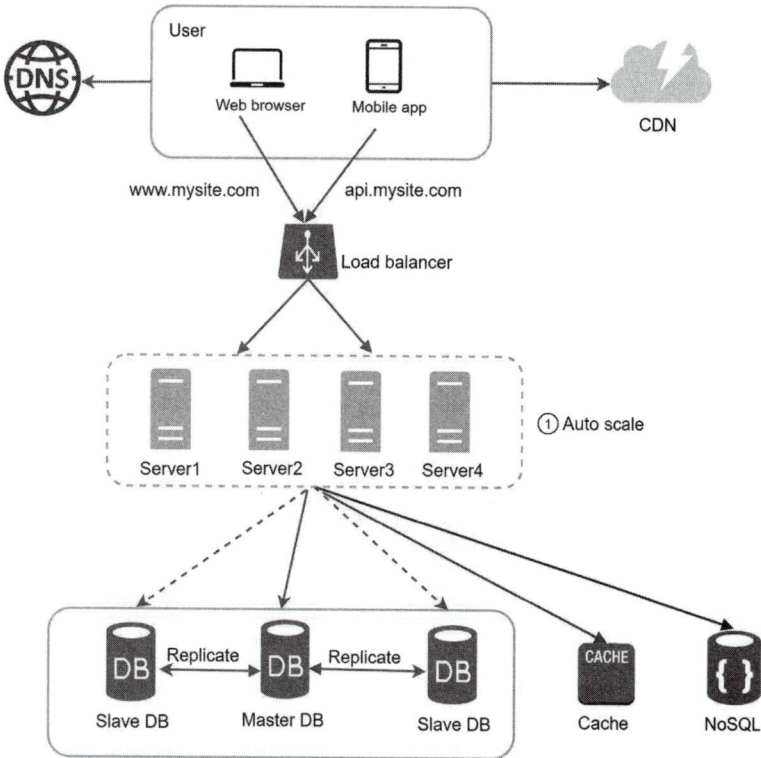

Figure 1-14

In Figure 1-14, we move the session data out of the web tier and store them in the persistent data store. The shared data store could be a relational database, Memcached/Redis, NoSQL, etc. The NoSQL data store is chosen as it is easy to scale. Autoscaling means adding or removing web servers automatically based on the traffic load. After the state data is removed out of web servers, auto-scaling of the web tier is easily achieved by adding or removing servers based on traffic load.

Your website grows rapidly and attracts a significant number of users internationally. To improve availability and provide a better user experience across wider geographical areas, supporting multiple data centers is crucial.

Data centers

Figure 1-15 shows an example setup with two data centers. In normal operation, users are geoDNS-routed, also known as geo-routed, to the closest data center, with a split traffic of $x\%$ in US-East and $(100 - x)\%$ in US-West. geoDNS is a DNS service that allows domain names to be resolved to IP addresses based on the location of a user.

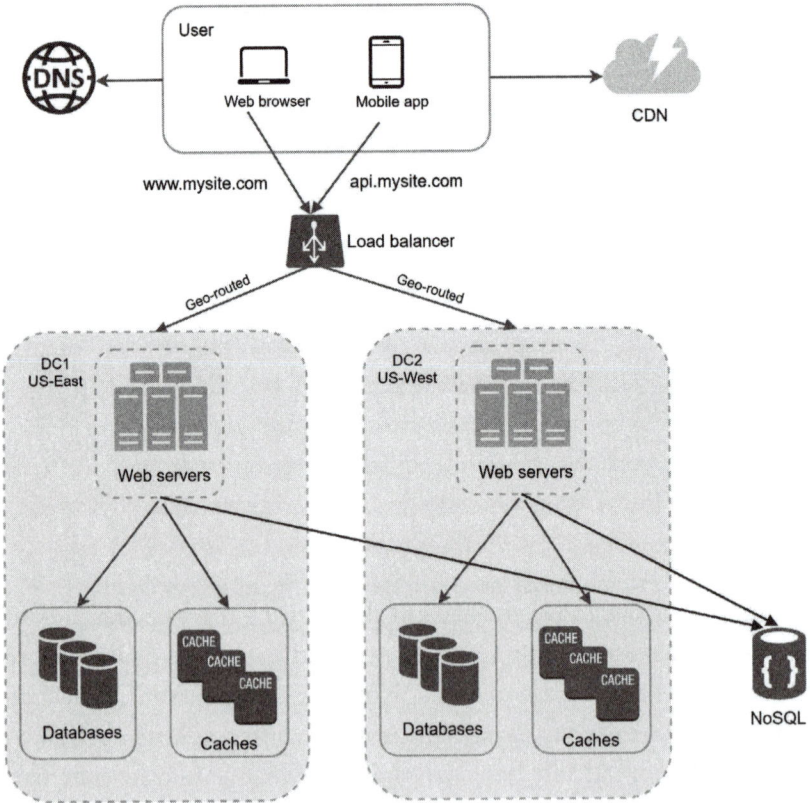

Figure 1-15

In the event of any significant data center outage, we direct all traffic to a healthy data center. In Figure 1-16, data center 2 (US-West) is offline, and 100% of the traffic is routed to data center 1 (US-East).

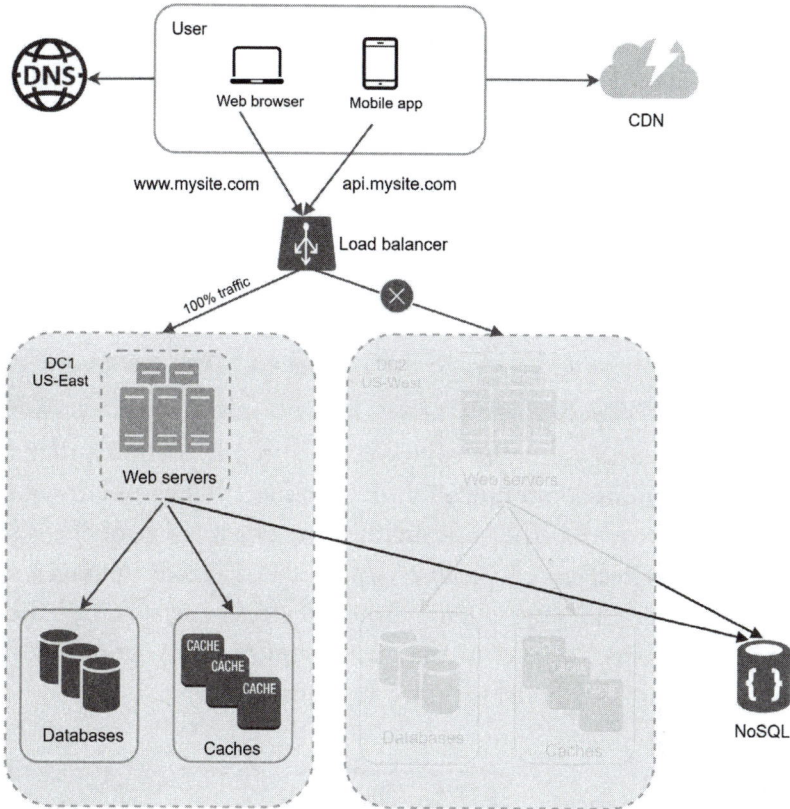

Figure 1-16

Several technical challenges must be resolved to achieve multi-data center setup:

- Traffic redirection: Effective tools are needed to direct traffic to the correct data center. GeoDNS can be used to direct traffic to the nearest data center depending on where a user is located.

- Data synchronization: Users from different regions could use different local databases or caches. In failover cases, traffic might be routed to a data center where data is unavailable. A common strategy is to replicate data across multiple data centers. A previous study shows how Netflix implements asynchronous multi-data center replication [11].

- Test and deployment: With multi-data center setup, it is important to test your website/application at different locations. Automated deployment tools are vital to keep services consistent through all the data centers [11].

To further scale our system, we need to decouple different components of the system so they can be scaled independently. Messaging queue is a key strategy employed by many real-world distributed systems to solve this problem.

Message queue

A message queue is a durable component, stored in memory, that supports asynchronous communication. It serves as a buffer and distributes asynchronous requests. The basic architecture of a message queue is simple. Input services, called producers/publishers, create messages, and publish them to a message queue. Other services or servers, called consumers/subscribers, connect to the queue, and perform actions defined by the messages. The model is shown in Figure 1-17.

Figure 1-17

Decoupling makes the message queue a preferred architecture for building a scalable and reliable application. With the message queue, the producer can post a message to the queue when the consumer is unavailable to process it. The consumer can read messages from the queue even when the producer is unavailable.

Consider the following use case: your application supports photo customization, including cropping, sharpening, blurring, etc. Those customization tasks take time to complete. In Figure 1-18, web servers publish photo processing jobs to the message queue. Photo processing workers pick up jobs from the message queue and asynchronously perform photo customization tasks. The producer and the consumer can be scaled independently. When the size of the queue becomes large, more workers are added to reduce the processing time. However, if the queue is empty most of the time, the number of workers can be reduced.

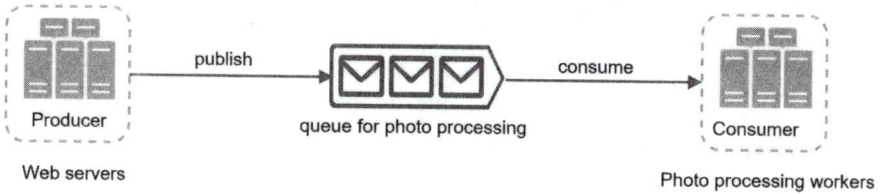

Figure 1-18

Logging, metrics, automation

When working with a small website that runs on a few servers, logging, metrics, and automation support are good practices but not a necessity. However, now that your site has grown to serve a large business, investing in those tools is essential.

Logging: Monitoring error logs is important because it helps to identify errors and problems in the system. You can monitor error logs at per server level or use tools to aggregate them to a centralized service for easy search and viewing.

Metrics: Collecting different types of metrics help us to gain business insights and understand the health status of the system. Some of the following metrics are useful:

- Host level metrics: CPU, Memory, disk I/O, etc.
- Aggregated level metrics: for example, the performance of the entire database tier, cache tier, etc.
- Key business metrics: daily active users, retention, revenue, etc.

Automation: When a system gets big and complex, we need to build or leverage automation tools to improve productivity. Continuous integration is a good practice, in which each code check-in is verified through automation, allowing teams to detect problems early. Besides, automating your build, test, deploy process, etc. could improve developer productivity significantly.

Adding message queues and different tools

Figure 1-19 shows the updated design. Due to the space constraint, only one data center is shown in the figure.

1. The design includes a message queue, which helps to make the system more loosely coupled and failure resilient.

2. Logging, monitoring, metrics, and automation tools are included.

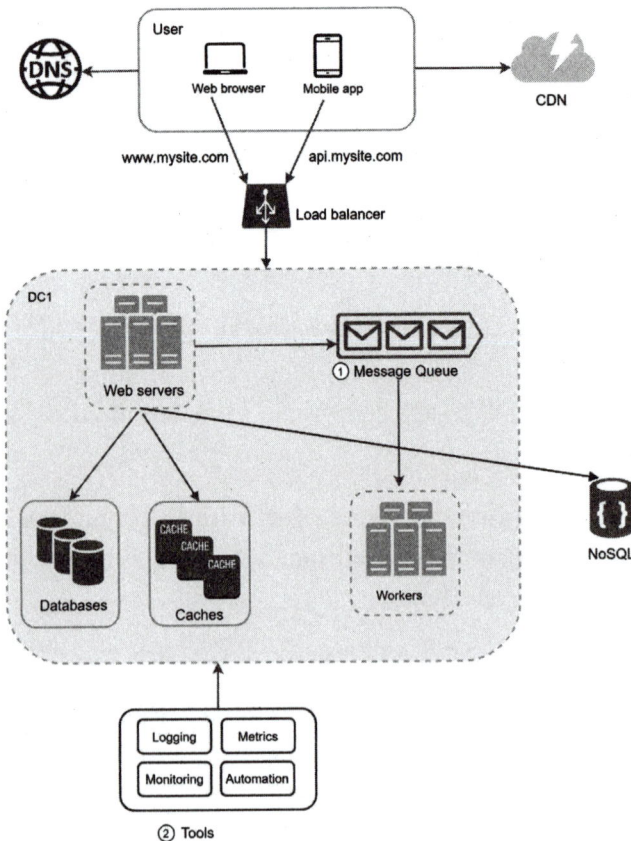

Figure 1-19

As the data grows every day, your database gets more overloaded. It is time to scale the data tier.

Database scaling

There are two broad approaches for database scaling: vertical scaling and horizontal scaling.

Vertical scaling

Vertical scaling, also known as scaling up, is the scaling by adding more power (CPU, RAM, DISK, etc.) to an existing machine. There are some powerful database servers. According to Amazon Relational Database Service (RDS) [12], you can get a database server with 24 TB of RAM. This kind of powerful database server could store and handle lots of data. For example, stackoverflow.com in 2013 had over 10 million monthly unique visitors, but it only had 1 master database [13]. However, vertical scaling comes with some serious drawbacks:

- You can add more CPU, RAM, etc. to your database server, but there are hardware limits. If you have a large user base, a single server is not enough.

- Greater risk of single point of failures.

- The overall cost of vertical scaling is high. Powerful servers are much more expensive.

Horizontal scaling

Horizontal scaling, also known as sharding, is the practice of adding more servers. Figure 1-20 compares vertical scaling with horizontal scaling.

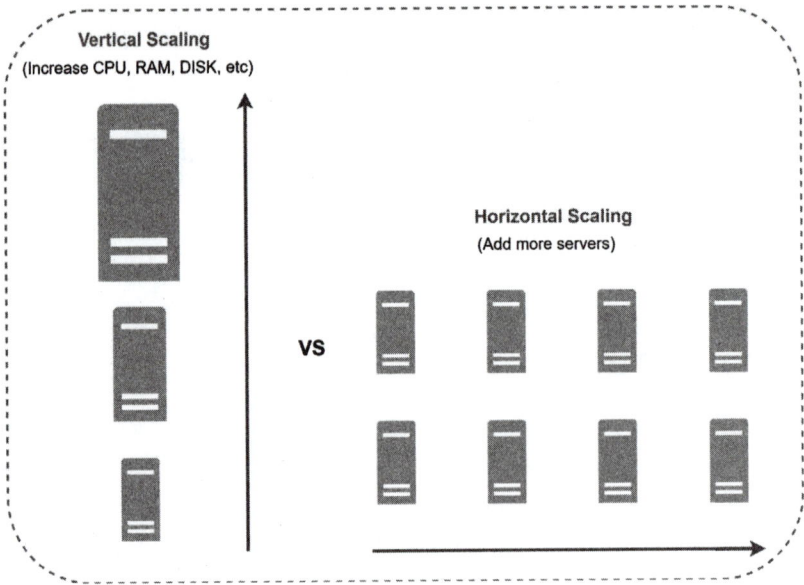

Figure 1-20

Sharding separates large databases into smaller, more easily managed parts called shards. Each shard shares the same schema, though the actual data on each shard is unique to the shard.

Figure 1-21 shows an example of sharded databases. User data is allocated to a database server based on user IDs. Anytime you access data, a hash function is used to find the corresponding shard. In our example, *user_id % 4* is used as the hash function. If the result equals to 0, shard 0 is used to store and fetch data. If the result equals to 1, shard 1 is used. The same logic applies to other shards.

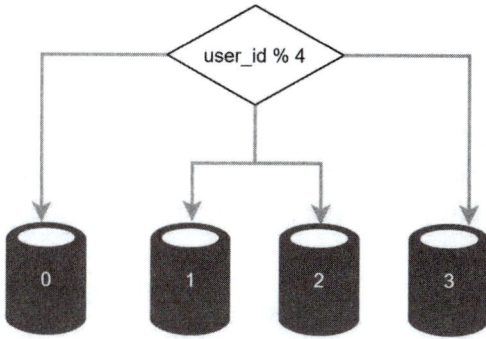

Figure 1-21

Figure 1-22 shows the user table in sharded databases.

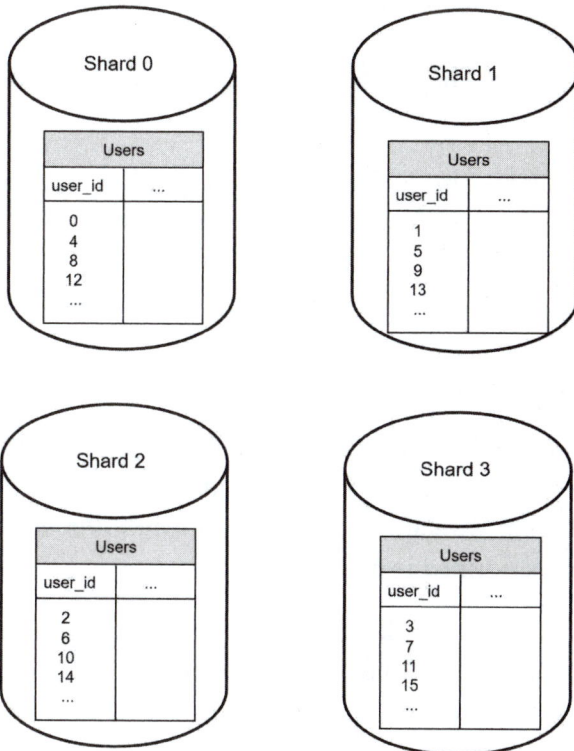

Figure 1-22

The most important factor to consider when implementing a sharding strategy is the choice of the sharding key. Sharding key (known as a partition key) consists of one or more columns that determine how data is distributed. As shown in Figure 1-22, *"user_id"* is the sharding key. A sharding key allows you to retrieve and modify data efficiently by routing database queries to the correct database. When choosing a sharding key, one of the most important criteria is to choose a key that can evenly distributed data.

Sharding is a great technique to scale the database but it is far from a perfect solution. It introduces complexities and new challenges to the system:

Resharding data: Resharding data is needed when 1) a single shard could no longer hold more data due to rapid growth. 2) Certain shards might experience shard exhaustion faster than others due to uneven data distribution. When shard exhaustion happens, it requires updating the sharding function and moving data around. Consistent hashing, which will be discussed in Chapter 5, is a commonly used technique to solve this problem.

Celebrity problem: This is also called a hotspot key problem. Excessive access to a specific shard could cause server overload. Imagine data for Katy Perry, Justin Bieber, and Lady Gaga all end up on the same shard. For social applications, that shard will be overwhelmed with read operations. To solve this problem, we may need to allocate a shard for each celebrity. Each shard might even require further partition.

Join and de-normalization: Once a database has been sharded across multiple servers, it is hard to perform join operations across database shards. A common workaround is to de-normalize the database so that queries can be performed in a single table.

In Figure 1-23, we shard databases to support rapidly increasing data traffic. At the same time, some of the non-relational functionalities are moved to a NoSQL data store to reduce the database load. Here is an article that covers many use cases of NoSQL [14].

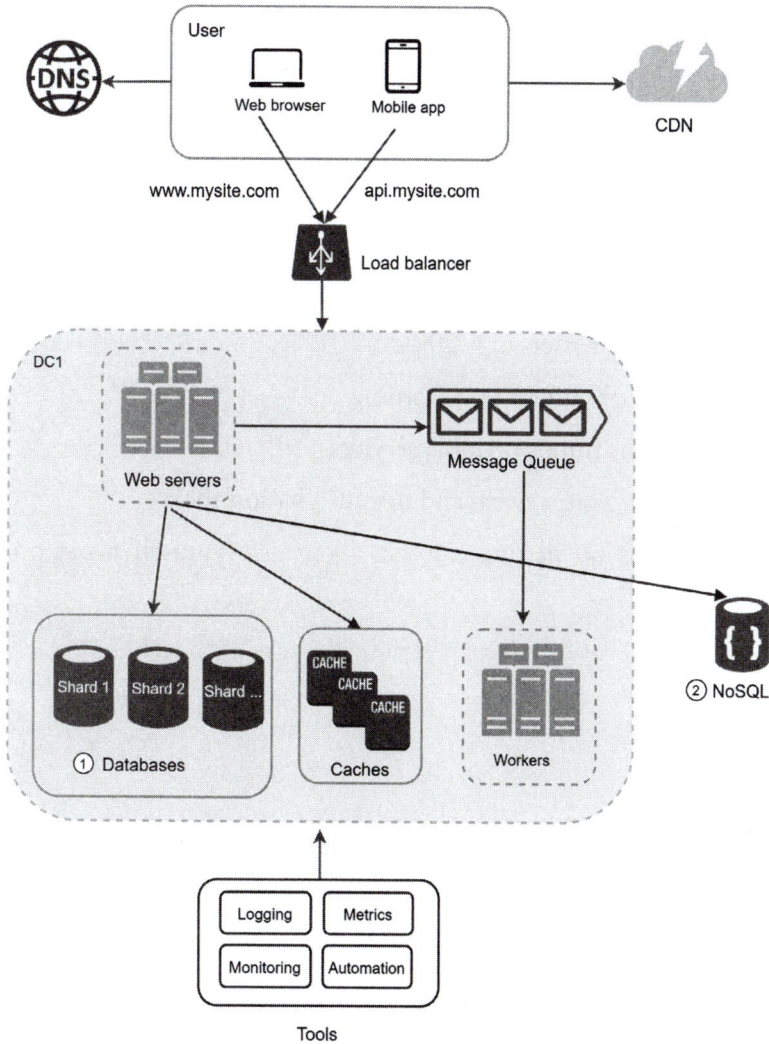

Figure 1-23

Millions of users and beyond

Scaling a system is an iterative process. Iterating on what we have learned in this chapter could get us far. More fine-tuning and new strategies are needed to scale beyond millions of users. For example, you might need to optimize your system and decouple the system to even smaller services.

All the techniques learned in this chapter should provide a good foundation to tackle new challenges. To conclude this chapter, we provide a summary of how we scale our system to support millions of users:

- Keep web tier stateless
- Build redundancy at every tier
- Cache data as much as you can
- Support multiple data centers
- Host static assets in CDN
- Scale your data tier by sharding
- Split tiers into individual services
- Monitor your system and use automation tools

Congratulations on getting this far! Now give yourself a pat on the back. Good job!

Reference materials

[1] Hypertext Transfer Protocol:
https://en.wikipedia.org/wiki/Hypertext_Transfer_Protocol

[2] Should you go Beyond Relational Databases?:
https://blog.teamtreehouse.com/should-you-go-beyond-
relational-databases

[3] Replication: https://en.wikipedia.org/wiki/Replication_
(computing)

[4] Multi-master replication:
https://en.wikipedia.org/wiki/Multi-master_replication

[5] NDB Cluster Replication: Multi-Master and Circular Replication:
https://dev.mysql.com/doc/refman/5.7/en/mysql-cluster-replication-
multi-master.html

[6] Caching Strategies and How to Choose the Right One:
https://codeahoy.com/2017/08/11/caching-strategies-and-how-to-
choose-the-right-one/

[7] R. Nishtala, "Facebook, Scaling Memcache at," 10th USENIX
Symposium on Networked Systems Design and Implementation
(NSDI '13).

[8] Single point of failure: https://en.wikipedia.org/wiki/Single_point_
of_failure

[9] Amazon CloudFront Dynamic Content Delivery:
https://aws.amazon.com/cloudfront/dynamic-content/

[10] Configure Sticky Sessions for Your Classic Load Balancer:
https://docs.aws.amazon.com/elasticloadbalancing/latest/classic/elb-
sticky-sessions.html

[11] Active-Active for Multi-Regional Resiliency:
https://netflixtechblog.com/active-active-for-multi-regional-resiliency-
c47719f6685b

[12] Amazon EC2 High Memory Instances:
https://aws.amazon.com/ec2/instance-types/high-memory/

[13] What it takes to run Stack Overflow:
http://nickcraver.com/blog/2013/11/22/what-it-takes-to-run-
stack-overflow

[14] What The Heck Are You Actually Using NoSQL For:
http://highscalability.com/blog/2010/12/6/what-the-heck-are-you-
actually-using-nosql-for.html

2

BACK-OF-THE-ENVELOPE ESTIMATION

In a system design interview, sometimes you are asked to estimate system capacity or performance requirements using a back-of-the-envelope estimation. According to Jeff Dean, Google Senior Fellow, "back-of-the-envelope calculations are estimates you create using a combination of thought experiments and common performance numbers to get a good feel for which designs will meet your requirements" [1].

You need to have a good sense of scalability basics to effectively carry out back-of-the-envelope estimation. The following concepts should be well understood: power of two [2], latency numbers every programmer should know, and availability numbers.

Power of two

Although data volume can become enormous when dealing with distributed systems, calculation all boils down to the basics. To obtain correct calculations, it is critical to know the data volume unit using the power of 2. A byte is a sequence of 8 bits. An ASCII character uses one byte of memory (8 bits). Below is a table explaining the data volume unit (Table 2-1).

Power	Approximate value	Full name	Short name
10	1 Thousand	1 Kilobyte	1 KB
20	1 Million	1 Megabyte	1 MB
30	1 Billion	1 Gigabyte	1 GB
40	1 Trillion	1 Terabyte	1 TB
50	1 Quadrillion	1 Petabyte	1 PB

Table 2-1

Latency numbers every programmer should know

Dr. Dean from Google reveals the length of typical computer operations in 2010 [1]. Some numbers are outdated as computers become faster and more powerful. However, those numbers should still be able to give us an idea of the fastness and slowness of different computer operations.

Operation name	Time
L1 cache reference	0.5 ns
Branch mispredict	5 ns
L2 cache reference	7 ns
Mutex lock/unlock	100 ns
Main memory reference	100 ns
Compress 1K bytes with Zippy	10,000 ns = 10 µs
Send 2K bytes over 1 Gbps network	20,000 ns = 20 µs
Read 1 MB sequentially from memory	250,000 ns = 250 µs
Round trip within the same datacenter	500,000 ns = 500 µs
Disk seek	10,000,000 ns = 10 ms
Read 1 MB sequentially from the network	10,000,000 ns = 10 ms
Read 1 MB sequentially from disk	30,000,000 ns = 30 ms
Send packet CA (California) ->Netherlands->CA	150,000,000 ns = 150 ms

Table 2-2

Notes

----

ns = nanosecond, µs = microsecond, ms = millisecond

1 ns = 10^-9 seconds

1 µs= 10^-6 seconds = 1,000 ns

1 ms = 10^-3 seconds = 1,000 μs = 1,000,000 ns

A Google software engineer built a tool to visualize Dr. Dean's numbers. The tool also takes the time factor into consideration. Figures 2-1 shows the visualized latency numbers as of 2020 (source of figures: reference material [3]).

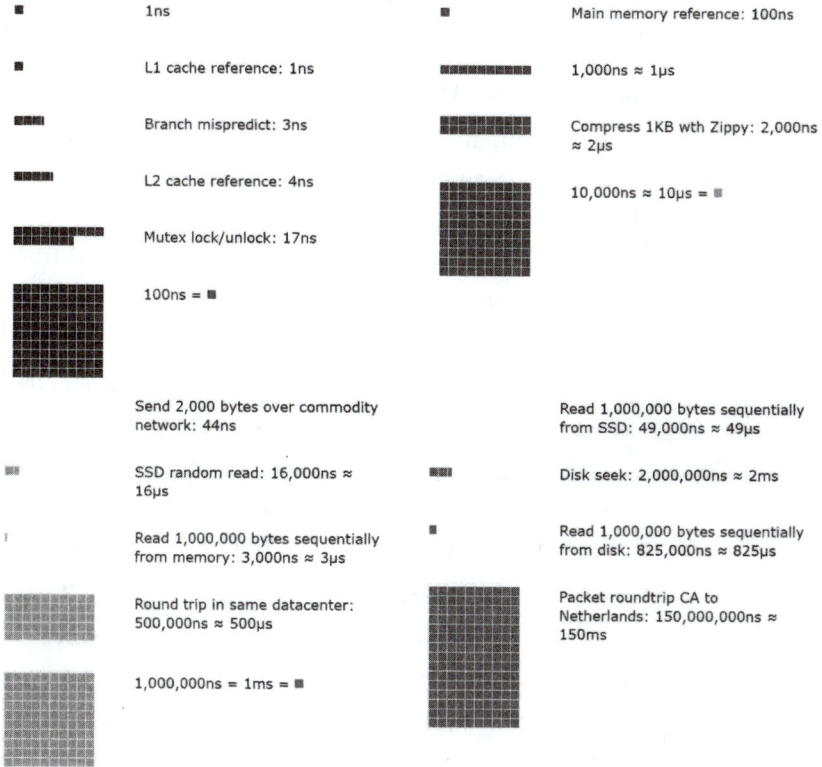

■	1ns	■	Main memory reference: 100ns
■	L1 cache reference: 1ns	■■■■■■■■■■	1,000ns ≈ 1μs
■■■■	Branch mispredict: 3ns	■■■■■■■■■	Compress 1KB wth Zippy: 2,000ns ≈ 2μs
■■■■■	L2 cache reference: 4ns		10,000ns ≈ 10μs = ■
■■■■■■■■■	Mutex lock/unlock: 17ns		
	100ns = ■		

Send 2,000 bytes over commodity network: 44ns

Read 1,000,000 bytes sequentially from SSD: 49,000ns ≈ 49μs

SSD random read: 16,000ns ≈ 16μs

Disk seek: 2,000,000ns ≈ 2ms

Read 1,000,000 bytes sequentially from memory: 3,000ns ≈ 3μs

Read 1,000,000 bytes sequentially from disk: 825,000ns ≈ 825μs

Round trip in same datacenter: 500,000ns ≈ 500μs

Packet roundtrip CA to Netherlands: 150,000,000ns ≈ 150ms

1,000,000ns = 1ms = ■

Figure 2-1

By analyzing the numbers in Figure 2-1, we get the following conclusions:

- Memory is fast but the disk is slow.
- Avoid disk seeks if possible.
- Simple compression algorithms are fast.
- Compress data before sending it over the internet if possible.

- Data centers are usually in different regions, and it takes time to send data between them.

Availability numbers

High availability is the ability of a system to be continuously operational for a desirably long period of time. High availability is measured as a percentage, with 100% means a service that has 0 downtime. Most services fall between 99% and 100%.

A service level agreement (SLA) is a commonly used term for service providers. This is an agreement between you (the service provider) and your customer, and this agreement formally defines the level of uptime your service will deliver. Cloud providers Amazon [4], Google [5] and Microsoft [6] set their SLAs at 99.9% or above. Uptime is traditionally measured in nines. The more the nines, the better. As shown in Table 2-3, the number of nines correlate to the expected system downtime.

Availability %	Downtime per day	Downtime per week	Downtime per month	Downtime per year
99%	14.40 minutes	1.68 hours	7.31 hours	3.65 days
99.9%	1.44 minutes	10.08 minutes	43.83 minutes	8.77 hours
99.99%	8.64 seconds	1.01 minutes	4.38 minutes	52.60 minutes
99.999%	864.00 milliseconds	6.05 seconds	26.30 seconds	5.26 minutes
99.9999%	86.40 milliseconds	604.80 milliseconds	2.63 seconds	31.56 seconds

Table 2-3

Example: Estimate Twitter QPS and storage requirements

Please note the following numbers are for this exercise only as they are not real numbers from Twitter.

Assumptions:

- 300 million monthly active users.
- 50% of users use Twitter daily.
- Users post 2 tweets per day on average.
- 10% of tweets contain media.
- Data is stored for 5 years.

Estimations:

Query per second (QPS) estimate:

- Daily active users (DAU) = 300 million * 50% = 150 million
- Tweets QPS = 150 million * 2 tweets / 24 hour / 3600 sec-onds = ~3500
- Peak QPS = 2 * QPS = ~7000

We will only estimate media storage here.

- Average tweet size:
    - o   tweet_id   64 bytes
    - o   text   140 bytes
    - o   media   1 MB
- Media storage: 150 million * 2 * 10% * 1 MB = 30 TB per day
- 5-year media storage: 30 TB * 365 * 5 = ~55 PB

Tips

Back-of-the-envelope estimation is all about the process. Solving the problem is more important than obtaining results. Interviewers may test your problem-solving skills. Here are a few tips to follow:

- Rounding and Approximation. It is difficult to perform complicated math operations during the interview. For example, what is the result of "99987 / 9.1"? There is no need to spend valuable time to solve complicated math problems. Precision is not expected. Use round numbers and approximation to your advantage. The division question can be simplified as follows: "100,000 / 10".

- Write down your assumptions. It is a good idea to write down your assumptions to be referenced later.

- Label your units. When you write down "5", does it mean 5 KB or 5 MB? You might confuse yourself with this. Write down the units because "5 MB" helps to remove ambiguity.

- Commonly asked back-of-the-envelope estimations: QPS, peak QPS, storage, cache, number of servers, etc. You can practice these calculations when preparing for an interview. Practice makes perfect.

Congratulations on getting this far! Now give yourself a pat on the back. Good job!

Reference materials

[1] J. Dean.Google Pro Tip: Use Back-Of-The-Envelope-Calculations To Choose The Best Design:
http://highscalability.com/blog/2011/1/26/google-pro-tip-use-back-of-the-envelope-calculations-to-choo.html

[2] System design primer: https://github.com/donnemartin/system-design-primer

[3] Latency Numbers Every Programmer Should Know:
https://colin-scott.github.io/personal_website/research/interactive_latency.html

[4] Amazon Compute Service Level Agreement:
https://aws.amazon.com/compute/sla/

[5] Compute Engine Service Level Agreement (SLA):
https://cloud.google.com/compute/sla

[6] SLA summary for Azure services:
https://azure.microsoft.com/en-us/support/legal/sla/summary/

A FRAMEWORK FOR SYSTEM DESIGN INTERVIEWS

You have just landed a coveted on-site interview at your dream company. The hiring coordinator sends you a schedule for that day. Scanning down the list, you feel pretty good about it until your eyes land on this interview session - System Design Interview.

System design interviews are often intimidating. It could be as vague as "designing a well-known product X?". The questions are ambiguous and seem unreasonably broad. Your weariness is understandable. After all, how could anyone design a popular product in an hour that has taken hundreds if not thousands of engineers to build?

The good news is that no one expects you to. Real-world system design is extremely complicated. For example, Google search is deceptively simple; however, the amount of technology that underpins that simplicity is truly astonishing. If no one expects you to design a real-world system in an hour, what is the benefit of a system design interview?

The system design interview simulates real-life problem solving where two co-workers collaborate on an ambiguous problem and come up with a solution that meets their goals. The problem is open-ended, and there is no perfect answer. The final design is less important compared to the work you put in the design process. This allows you to demonstrate your design skill, defend your design choices, and respond to feedback in a constructive manner.

Let us flip the table and consider what goes through the interviewer's head as she walks into the conference room to meet you. The primary goal of the interviewer is to accurately assess your abilities. The last thing she wants is to give an inconclusive evaluation because the session has gone poorly and there are not enough signals. What is an interviewer looking for in a system design interview?

Many think that system design interview is all about a person's technical design skills. It is much more than that. An effective system design interview gives strong signals about a person's ability to collaborate, to work under pressure, and to resolve ambiguity constructively. The ability to ask good questions is also an essential skill, and many interviewers specifically look for this skill.

A good interviewer also looks for red flags. Over-engineering is a real disease of many engineers as they delight in design purity and ignore tradeoffs. They are often unaware of the compounding costs of over-engineered systems, and many companies pay a high price for that ignorance. You certainly do not want to demonstrate this tendency in a system design interview. Other red flags include narrow mindedness, stubbornness, etc.

In this chapter, we will go over some useful tips and introduce a simple and effective framework to solve system design interview problems.

A 4-step process for effective system design interview

Every system design interview is different. A great system design interview is open-ended and there is no one-size-fits-all solution. However, there are steps and common ground to cover in every system design interview.

Step 1 - Understand the problem and establish design scope

"Why did the tiger roar?"

A hand shot up in the back of the class.

"Yes, Jimmy?", the teacher responded.

"Because he was HUNGRY".

"Very good Jimmy."

Throughout his childhood, Jimmy has always been the first to answer questions in the class. Whenever the teacher asks a question, there is always a kid in the classroom who loves to take a crack at the question, no matter if he knows the answer or not. That is Jimmy.

Jimmy is an ace student. He takes pride in knowing all the answers fast. In exams, he is usually the first person to finish the questions. He is a teacher's top choice for any academic competition.

DON'T be like Jimmy.

In a system design interview, giving out an answer quickly without thinking gives you no bonus points. Answering without a thorough understanding of the requirements is a huge red flag as the interview is not a trivia contest. There is no right answer.

So, do not jump right in to give a solution. Slow down. Think deeply and ask questions to clarify requirements and assumptions. This is extremely important.

As an engineer, we like to solve hard problems and jump into the final design; however, this approach is likely to lead you to design the wrong system. One of the most important skills as an engineer is to ask the right questions, make the proper assumptions, and gather all the information needed to build a system. So, do not be afraid to ask questions.

When you ask a question, the interviewer either answers your question directly or asks you to make your assumptions. If the latter happens, write down your assumptions on the whiteboard or paper. You might need them later.

What kind of questions to ask? Ask questions to understand the exact requirements. Here is a list of questions to help you get started:

- What specific features are we going to build?
- How many users does the product have?
- How fast does the company anticipate to scale up? What are the anticipated scales in 3 months, 6 months, and a year?

- What is the company's technology stack? What existing services you might leverage to simplify the design?

Example

If you are asked to design a news feed system, you want to ask questions that help you clarify the requirements. The conversation between you and the interviewer might look like this:

Candidate: Is this a mobile app? Or a web app? Or both?
Interviewer: Both.

Candidate: What are the most important features for the product?
Interviewer: Ability to make a post and see friends' news feed.

Candidate: Is the news feed sorted in reverse chronological order or a particular order? The particular order means each post is given a different weight. For instance, posts from your close friends are more important than posts from a group.
Interviewer: To keep things simple, let us assume the feed is sorted by reverse chronological order.

Candidate: How many friends can a user have?
Interviewer: 5000

Candidate: What is the traffic volume?
Interviewer: 10 million daily active users (DAU)

Candidate: Can feed contain images, videos, or just text?
Interviewer: It can contain media files, including both images and videos.

Above are some sample questions that you can ask your interviewer. It is important to understand the requirements and clarify ambiguities

Step 2 - Propose high-level design and get buy-in

In this step, we aim to develop a high-level design and reach an agreement with the interviewer on the design. It is a great idea to collaborate with the interviewer during the process.

- Come up with an initial blueprint for the design. Ask for feedback. Treat your interviewer as a teammate and work together. Many good interviewers love to talk and get involved.

- Draw box diagrams with key components on the whiteboard or paper. This might include clients (mobile/web), APIs, web servers, data stores, cache, CDN, message queue, etc.

- Do back-of-the-envelope calculations to evaluate if your blueprint fits the scale constraints. Think out loud. Communicate with your interviewer if back-of-the-envelope is necessary before diving into it.

If possible, go through a few concrete use cases. This will help you frame the high-level design. It is also likely that the use cases would help you discover edge cases you have not yet considered.

Should we include API endpoints and database schema here? This depends on the problem. For large design problems like "Design Google search engine", this is a bit of too low level. For a problem like designing the backend for a multi-player poker game, this is a fair game. Communicate with your interviewer.

Example

Let us use "Design a news feed system" to demonstrate how to approach the high-level design. Here you are not required to understand how the system actually works. All the details will be explained in Chapter 11.

At the high level, the design is divided into two flows: feed publishing and news feed building.

- Feed publishing: when a user publishes a post, corresponding data is written into cache/database, and the post will be populated into friends' news feed.

- Newsfeed building: the news feed is built by aggregating friends' posts in a reverse chronological order.

Figure 3-1 and Figure 3-2 present high-level designs for feed publishing and news feed building flows, respectively.

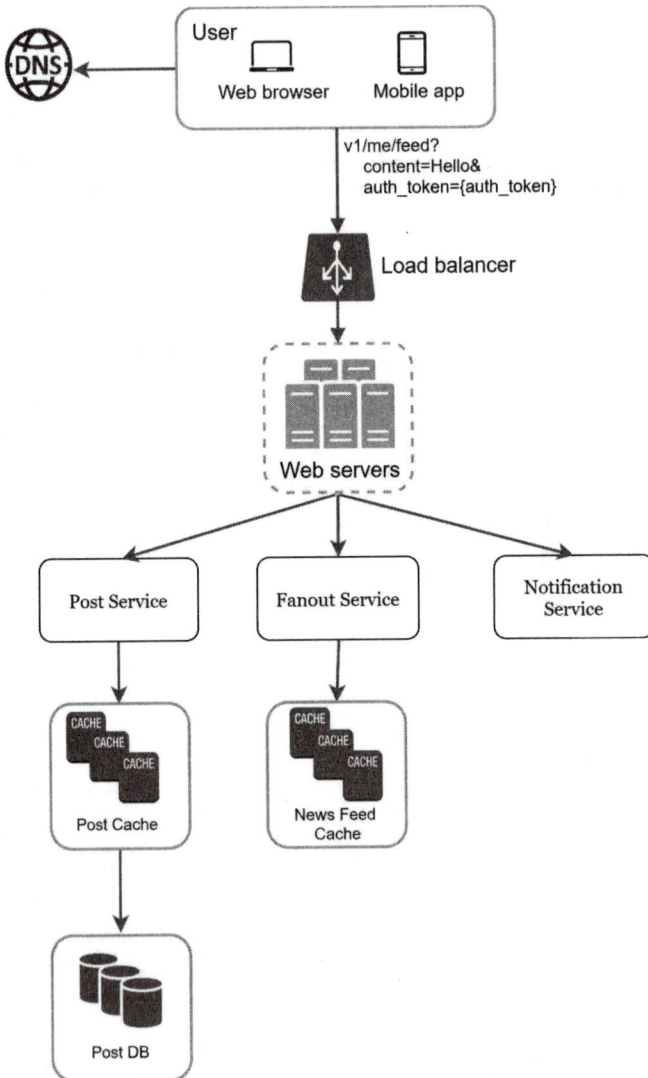

Figure 3-1

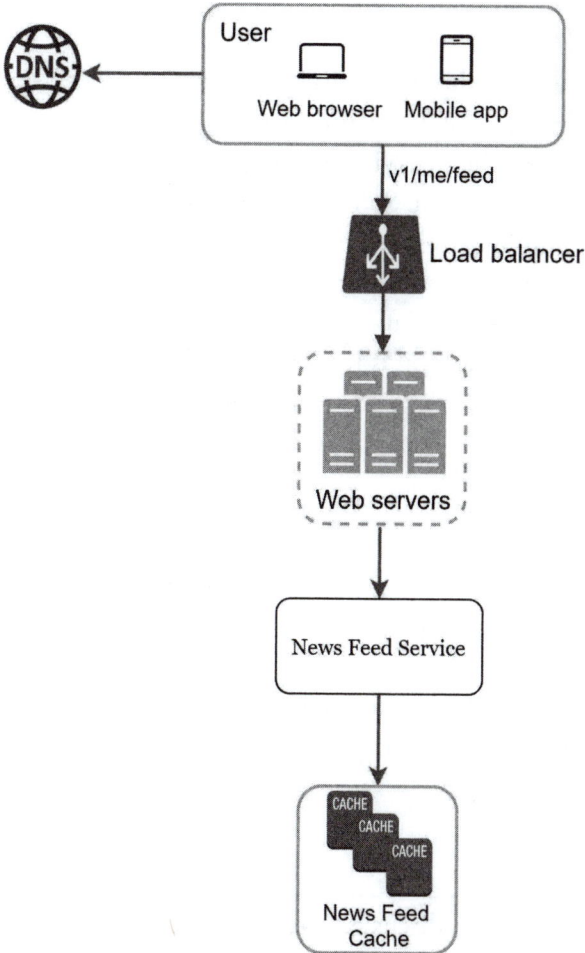

Figure 3-2

Step 3 - Design deep dive

At this step, you and your interviewer should have already achieved the following objectives:

- Agreed on the overall goals and feature scope
- Sketched out a high-level blueprint for the overall design
- Obtained feedback from your interviewer on the high-level design

- Had some initial ideas about areas to focus on in deep dive based on her feedback

You shall work with the interviewer to identify and prioritize components in the architecture. It is worth stressing that every interview is different. Sometimes, the interviewer may give off hints that she likes focusing on high-level design. Sometimes, for a senior candidate interview, the discussion could be on the system performance characteristics, likely focusing on the bottlenecks and resource estimations. In most cases, the interviewer may want you to dig into details of some system components. For URL shortener, it is interesting to dive into the hash function design that converts a long URL to a short one. For a chat system, how to reduce latency and how to support online/offline status are two interesting topics.

Time management is essential as it is easy to get carried away with minute details that do not demonstrate your abilities. You must be armed with signals to show your interviewer. Try not to get into unnecessary details. For example, talking about the EdgeRank algorithm of Facebook feed ranking in detail is not ideal during a system design interview as this takes much precious time and does not prove your ability in designing a scalable system.

Example

At this point, we have discussed the high-level design for a news feed system, and the interviewer is happy with your proposal. Next, we will investigate two of the most important use cases:

1. Feed publishing
2. News feed retrieval

Figure 3-3 and Figure 3-4 show the detailed design for the two use cases, which will be explained in detail in Chapter 11.

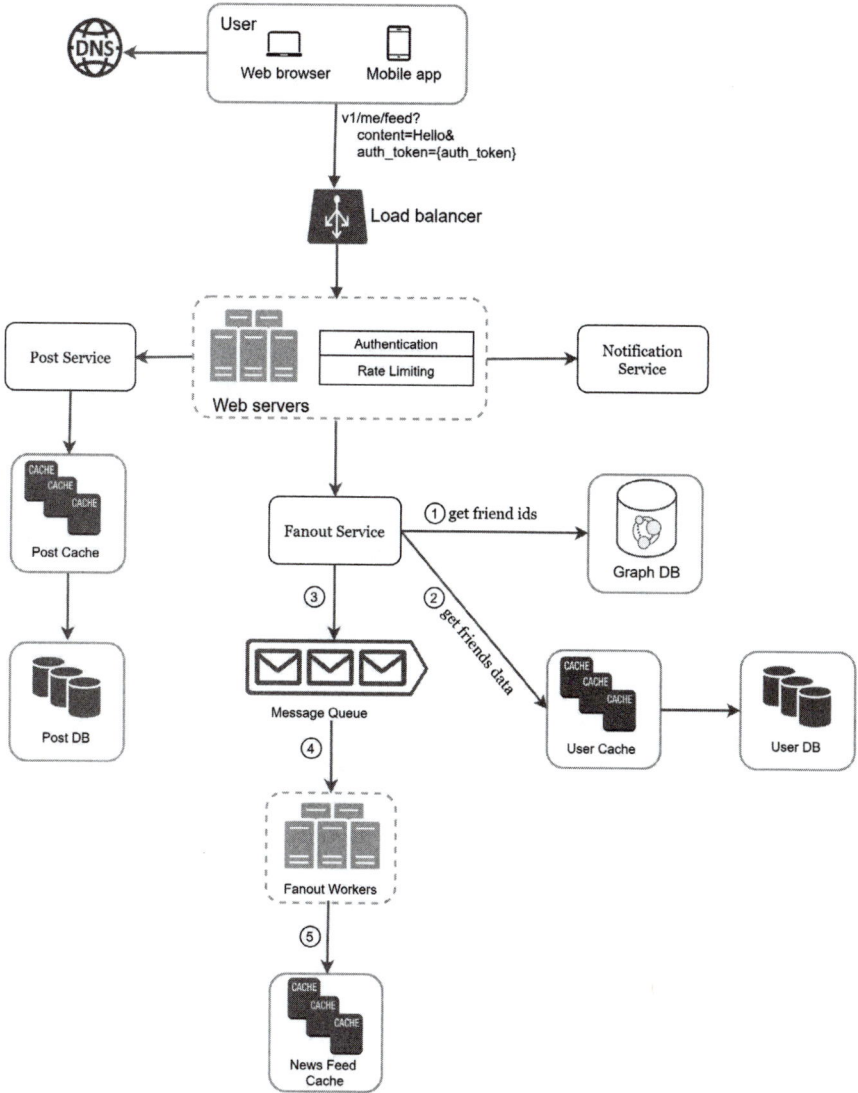

Figure 3-3

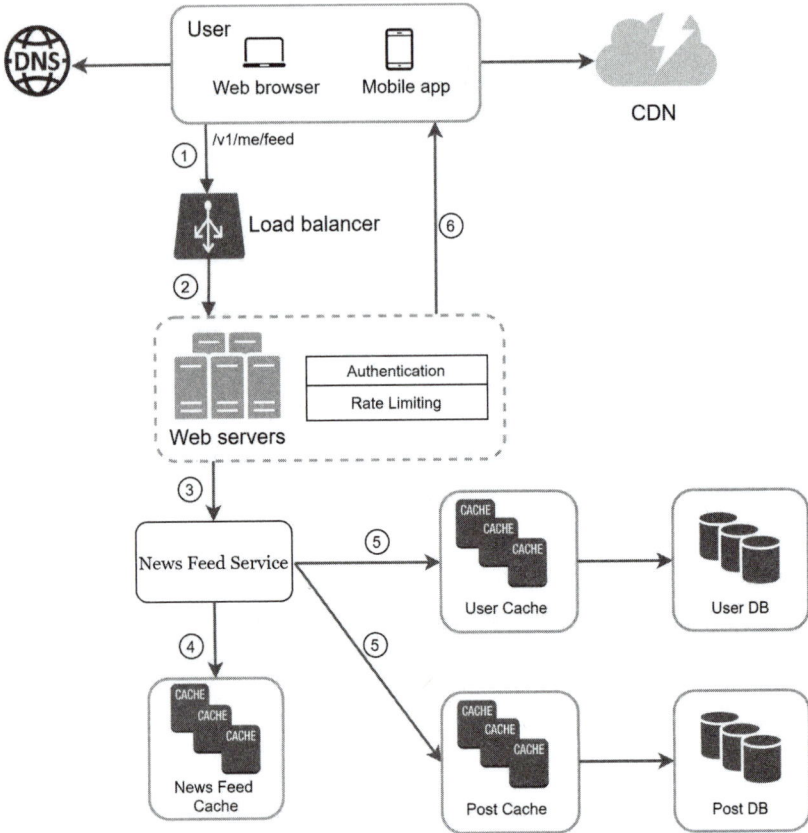

Figure 3-4

Step 4 - Wrap up

In this final step, the interviewer might ask you a few follow-up questions or give you the freedom to discuss other additional points. Here are a few directions to follow:

- The interviewer might want you to identify the system bottle-necks and discuss potential improvements. Never say your design is perfect and nothing can be improved. There is always some-thing to improve upon. This is a great opportunity to show your critical thinking and leave a good final impression.

- It could be useful to give the interviewer a recap of your design. This is particularly important if you suggested a few solutions. Refreshing your interviewer's memory can be helpful after a long session.

- Error cases (server failure, network loss, etc.) are interesting to talk about.

- Operation issues are worth mentioning. How do you monitor metrics and error logs? How to roll out the system?

- How to handle the next scale curve is also an interesting topic. For example, if your current design supports 1 million users, what changes do you need to make to support 10 million users?

- Propose other refinements you need if you had more time.

To wrap up, we summarize a list of the Dos and Don'ts.

Dos

- Always ask for clarification. Do not assume your assumption is correct.

- Understand the requirements of the problem.

- There is neither the right answer nor the best answer. A solution designed to solve the problems of a young startup is different from that of an established company with millions of users. Make sure you understand the requirements.

- Let the interviewer know what you are thinking. Communicate with your interview.

- Suggest multiple approaches if possible.

- Once you agree with your interviewer on the blueprint, go into details on each component. Design the most critical components first.

- Bounce ideas off the interviewer. A good interviewer works with you as a teammate.

- Never give up.

Don'ts

- Don't be unprepared for typical interview questions.

- Don't jump into a solution without clarifying the requirements and assumptions.

- Don't go into too much detail on a single component in the beginning. Give the high-level design first then drills down.

- If you get stuck, don't hesitate to ask for hints.

- Again, communicate. Don't think in silence.

- Don't think your interview is done once you give the design. You are not done until your interviewer says you are done. Ask for feedback early and often.

Time allocation on each step

System design interview questions are usually very broad, and 45 minutes or an hour is not enough to cover the entire design. Time management is essential. How much time should you spend on each step? The following is a very rough guide on distributing your time in a 45-minute interview session. Please remember this is a rough estimate, and the actual time distribution depends on the scope of the problem and the requirements from the interviewer.

Step 1 Understand the problem and establish design scope: 3 - 10 minutes

Step 2 Propose high-level design and get buy-in: 10 - 15 minutes

Step 3 Design deep dive: 10 - 25 minutes

Step 4 Wrap: 3 - 5 minutes

4

DESIGN A RATE LIMITER

In a network system, a rate limiter is used to control the rate of traffic sent by a client or a service. In the HTTP world, a rate limiter limits the number of client requests allowed to be sent over a specified period. If the API request count exceeds the threshold defined by the rate limiter, all the excess calls are blocked. Here are a few examples:

- A user can write no more than 2 posts per second.

- You can create a maximum of 10 accounts per day from the same IP address.

- You can claim rewards no more than 5 times per week from the same device.

In this chapter, you are asked to design a rate limiter. Before starting the design, we first look at the benefits of using an API rate limiter:

- Prevent resource starvation caused by Denial of Service (DoS) attack [1]. Almost all APIs published by large tech companies enforce some form of rate limiting. For example, Twitter limits the number of tweets to 300 per 3 hours [2]. Google docs APIs have the following default limit: 300 per user per 60 seconds for read requests [3]. A rate limiter prevents DoS attacks, either intentional or unintentional, by blocking the excess calls.

- Reduce cost. Limiting excess requests means fewer servers and allocating more resources to high priority APIs. Rate limiting is extremely important for companies that use paid third party APIs. For example, you are charged on a per-call basis for the following external APIs: check credit, make a payment, retrieve health records, etc. Limiting the number of calls is essential to reduce costs.

- Prevent servers from being overloaded. To reduce server load, a rate limiter is used to filter out excess requests caused by bots or users' misbehavior.

Step 1 - Understand the problem and establish design scope

Rate limiting can be implemented using different algorithms, each with its pros and cons. The interactions between an interviewer and a candidate help to clarify the type of rate limiters we are trying to build.

Candidate: What kind of rate limiter are we going to design? Is it a client-side rate limiter or server-side API rate limiter?
Interviewer: Great question. We focus on the server-side API rate limiter.

Candidate: Does the rate limiter throttle API requests based on IP, the user ID, or other properties?
Interviewer: The rate limiter should be flexible enough to support different sets of throttle rules.

Candidate: What is the scale of the system? Is it built for a startup or a big company with a large user base?
Interviewer: The system must be able to handle a large number of requests.

Candidate: Will the system work in a distributed environment?
Interviewer: Yes.

Candidate: Is the rate limiter a separate service or should it be implemented in application code?
Interviewer: It is a design decision up to you.

Candidate: Do we need to inform users who are throttled?
Interviewer: Yes.

Requirements

Here is a summary of the requirements for the system:

- Accurately limit excessive requests.

- Low latency. The rate limiter should not slow down HTTP response time.

- Use as little memory as possible.

- Distributed rate limiting. The rate limiter can be shared across multiple servers or processes.

- Exception handling. Show clear exceptions to users when their requests are throttled.

- High fault tolerance. If there are any problems with the rate limiter (for example, a cache server goes offline), it does not affect the entire system.

Step 2 - Propose high-level design and get buy-in

Let us keep things simple and use a basic client and server model for communication.

Where to put the rate limiter?

Intuitively, you can implement a rate limiter at either the client or server-side.

- Client-side implementation. Generally speaking, client is an unreliable place to enforce rate limiting because client requests can easily be forged by malicious actors. Moreover, we might not have control over the client implementation.

- Server-side implementation. Figure 4-1 shows a rate limiter that is placed on the server-side.

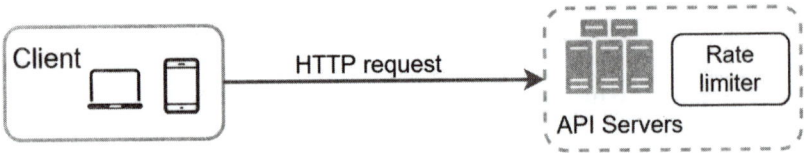

Figure 4-1

Besides the client and server-side implementations, there is an alternative way. Instead of putting a rate limiter at the API servers, we create a rate limiter middleware, which throttles requests to your APIs as shown in Figure 4-2.

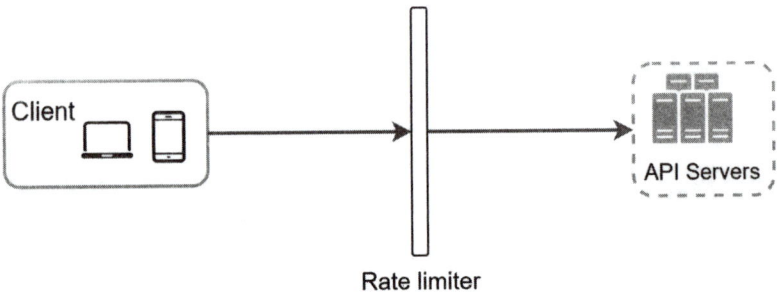

Figure 4-2

Let us use an example in Figure 4-3 to illustrate how rate limiting works in this design. Assume our API allows 2 requests per second, and a client sends 3 requests to the server within a second. The first two requests are routed to API servers. However, the rate limiter middleware throttles the third request and returns a HTTP status code 429. The HTTP 429 response status code indicates a user has sent too many requests.

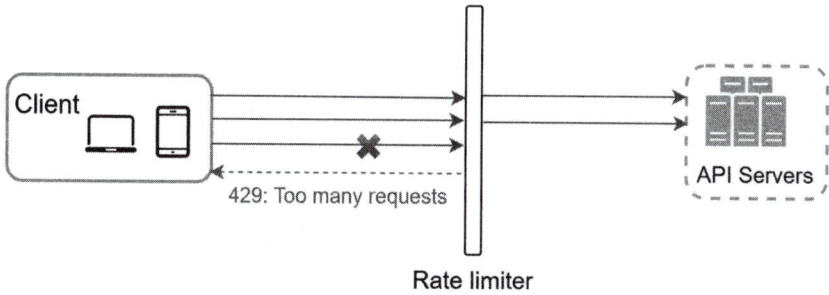

Figure 4-3

Cloud microservices [4] have become widely popular and rate limiting is usually implemented within a component called API gateway. API gateway is a fully managed service that supports rate limiting, SSL termination, authentication, IP whitelisting, servicing static content, etc. For now, we only need to know that the API gateway is a middleware that supports rate limiting.

While designing a rate limiter, an important question to ask ourselves is: where should the rater limiter be implemented, on the server-side or in a gateway? There is no absolute answer. It depends on your company's current technology stack, engineering resources, priorities, goals, etc. Here are a few general guidelines:

- Evaluate your current technology stack, such as programming language, cache service, etc. Make sure your current programming language is efficient to implement rate limiting on the server-side.

- Identify the rate limiting algorithm that fits your business needs. When you implement everything on the server-side, you have full control of the algorithm. However, your choice might be limited if you use a third-party gateway.

- If you have already used microservice architecture and included an API gateway in the design to perform authentication, IP whitelisting, etc., you may add a rate limiter to the API gateway.

- Building your own rate limiting service takes time. If you do not have enough engineering resources to implement a rate limiter, a commercial API gateway is a better option.

Algorithms for rate limiting

Rate limiting can be implemented using different algorithms, and each of them has distinct pros and cons. Even though this chapter does not focus on algorithms, understanding them at high-level helps to choose the right algorithm or combination of algorithms to fit our use cases. Here is a list of popular algorithms:

- Token bucket
- Leaking bucket
- Fixed window counter
- Sliding window log
- Sliding window counter

Token bucket algorithm

The token bucket algorithm is widely used for rate limiting. It is simple, well understood and commonly used by internet companies. Both Amazon [5] and Stripe [6] use this algorithm to throttle their API requests.

The token bucket algorithm work as follows:

- A token bucket is a container that has pre-defined capacity. Tokens are put in the bucket at preset rates periodically. Once the bucket is full, no more tokens are added. As shown in Figure 4-4, the token bucket capacity is 4. The refiller puts 2 tokens into the bucket every second. Once the bucket is full, extra tokens will overflow.

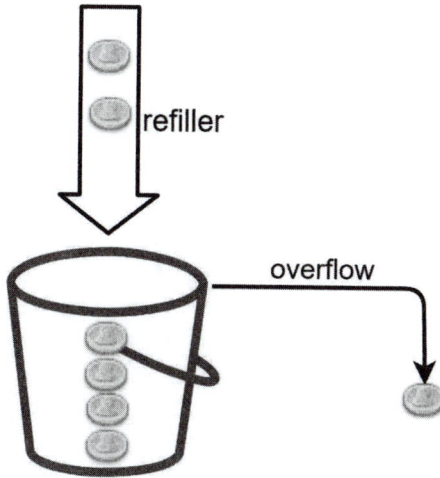

Figure 4-4

- Each request consumes one token. When a request arrives, we check if there are enough tokens in the bucket. Figure 4-5 explains how it works.

  o   If there are enough tokens, we take one token out for each request, and the request goes through.

  o   If there are not enough tokens, the request is dropped.

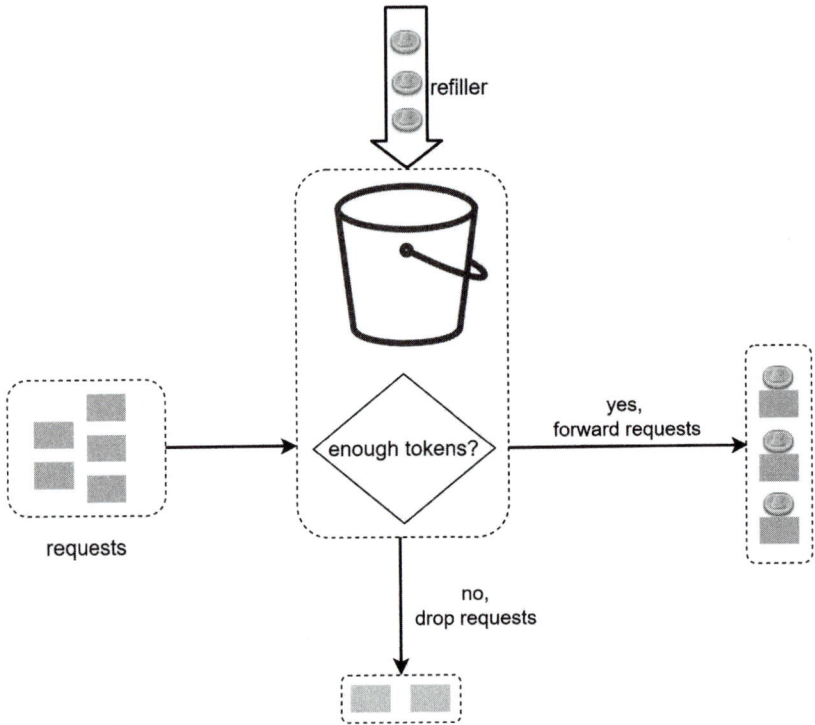

Figure 4-5

Figure 4-6 illustrates how token consumption, refill, and rate limiting logic work. In this example, the token bucket size is 4, and the refill rate is 4 per 1 minute.

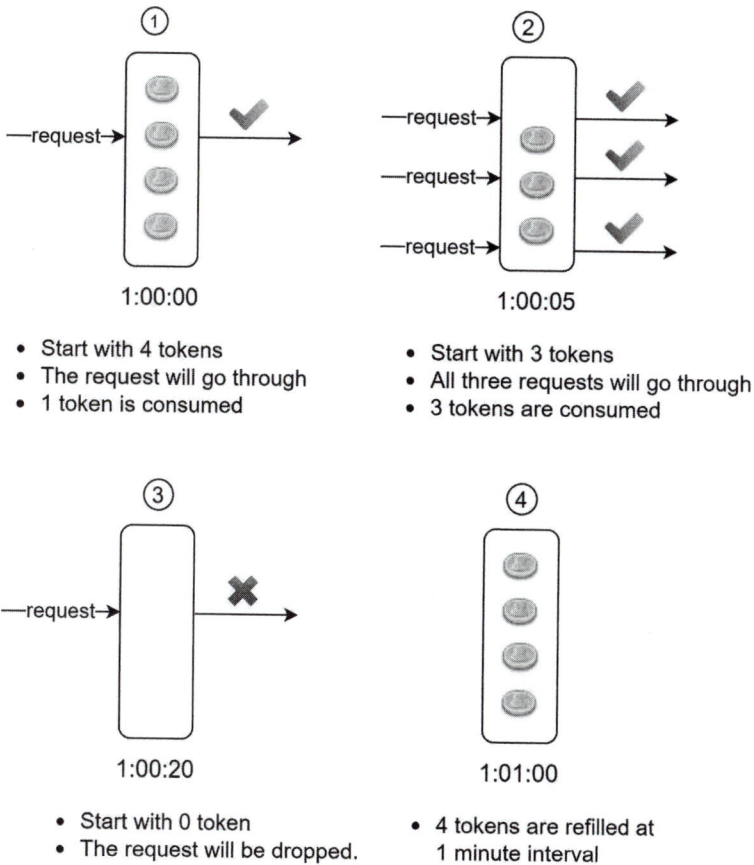

Figure 4-6

The token bucket algorithm takes two parameters:

- Bucket size: the maximum number of tokens allowed in the bucket
- Refill rate: number of tokens put into the bucket every second

How many buckets do we need? This varies, and it depends on the rate-limiting rules. Here are a few examples.

- It is usually necessary to have different buckets for different API endpoints. For instance, if a user is allowed to make 1 post per second, add 150 friends per day, and like 5 posts per second, 3 buckets are required for each user.

- If we need to throttle requests based on IP addresses, each IP address requires a bucket.

- If the system allows a maximum of 10,000 requests per second, it makes sense to have a global bucket shared by all requests.

Pros:

- The algorithm is easy to implement.

- Memory efficient.

- Token bucket allows a burst of traffic for short periods. A request can go through as long as there are tokens left.

Cons:

- Two parameters in the algorithm are bucket size and token refill rate. However, it might be challenging to tune them properly.

Leaking bucket algorithm

The leaking bucket algorithm is similar to the token bucket except that requests are processed at a fixed rate. It is usually implemented with a first-in-first-out (FIFO) queue. The algorithm works as follows:

- When a request arrives, the system checks if the queue is full. If it is not full, the request is added to the queue.

- Otherwise, the request is dropped.

- Requests are pulled from the queue and processed at regular intervals.

Figure 4-7 explains how the algorithm works.

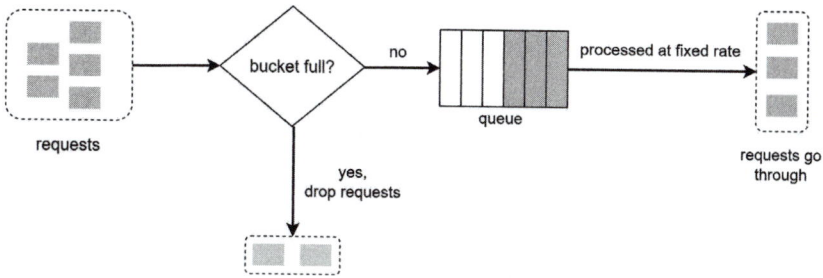

Figure 4-7

Leaking bucket algorithm takes the following two parameters:

- Bucket size: it is equal to the queue size. The queue holds the requests to be processed at a fixed rate.

- Outflow rate: it defines how many requests can be processed at a fixed rate, usually in seconds.

Shopify, an ecommerce company, uses leaky buckets for rate-limiting [7].

Pros:

- Memory efficient given the limited queue size.

- Requests are processed at a fixed rate therefore it is suitable for use cases that a stable outflow rate is needed.

Cons:

- A burst of traffic fills up the queue with old requests, and if they are not processed in time, recent requests will be rate limited.

- There are two parameters in the algorithm. It might not be easy to tune them properly.

Fixed window counter algorithm

Fixed window counter algorithm works as follows:

- The algorithm divides the timeline into fix-sized time windows and assign a counter for each window.
- Each request increments the counter by one.
- Once the counter reaches the pre-defined threshold, new requests are dropped until a new time window starts.

Let us use a concrete example to see how it works. In Figure 4-8, the time unit is 1 second and the system allows a maximum of 3 requests per second. In each second window, if more than 3 requests are received, extra requests are dropped as shown in Figure 4-8.

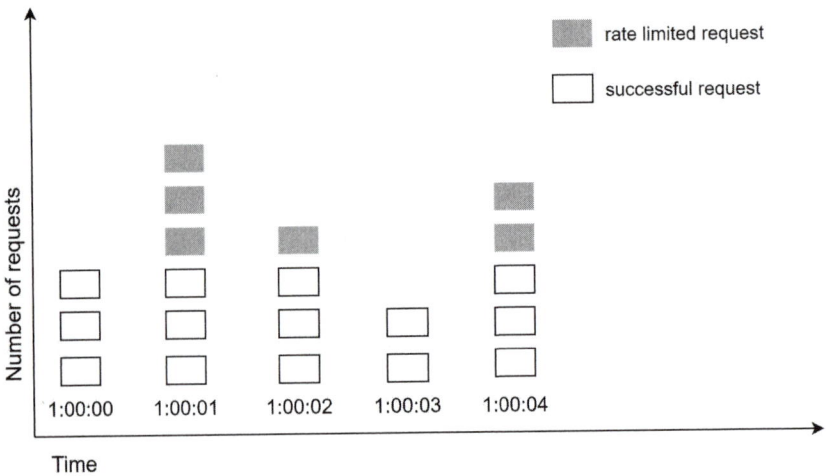

Figure 4-8

A major problem with this algorithm is that a burst of traffic at the edges of time windows could cause more requests than allowed quota to go through. Consider the following case:

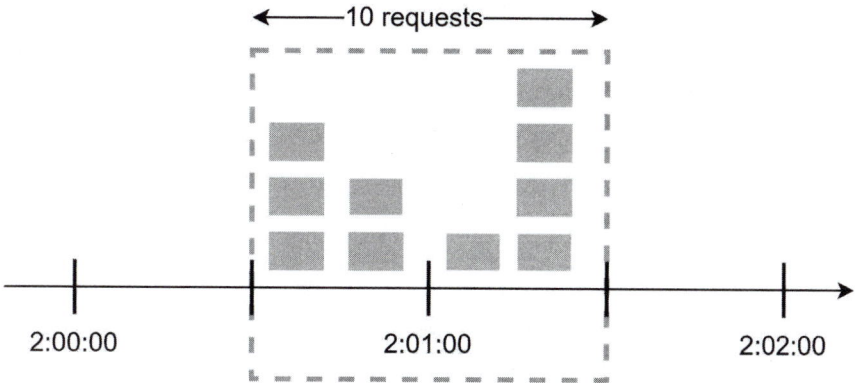

Figure 4-9

In Figure 4-9, the system allows a maximum of 5 requests per minute, and the available quota resets at the human-friendly round minute. As seen, there are five requests between 2:00:00 and 2:01:00 and five more requests between 2:01:00 and 2:02:00. For the one-minute window between 2:00:30 and 2:01:30, 10 requests go through. That is twice as many as allowed requests.

Pros:

- Memory efficient.

- Easy to understand.

- Resetting available quota at the end of a unit time window fits certain use cases.

Cons:

- Spike in traffic at the edges of a window could cause more requests than the allowed quota to go through.

Sliding window log algorithm

As discussed previously, the fixed window counter algorithm has a major issue: it allows more requests to go through at the edges of a window. The sliding window log algorithm fixes the issue. It works as follows:

- The algorithm keeps track of request timestamps. Timestamp data is usually kept in cache, such as sorted sets of Redis [8].

- When a new request comes in, remove all the outdated time-stamps. Outdated timestamps are defined as those older than the start of the current time window.

- Add timestamp of the new request to the log.

- If the log size is the same or lower than the allowed count, a request is accepted. Otherwise, it is rejected.

We explain the algorithm with an example as revealed in Figure 4-10.

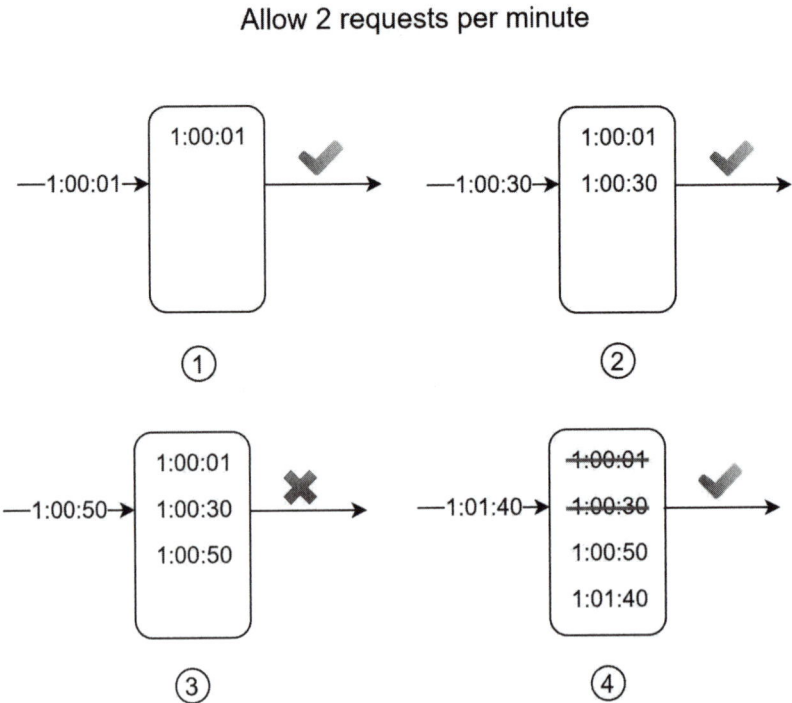

Figure 4-10

In this example, the rate limiter allows 2 requests per minute. Usually, Linux timestamps are stored in the log. However, human-readable representation of time is used in our example for better readability.

- The log is empty when a new request arrives at 1:00:01. Thus, the request is allowed.

- A new request arrives at 1:00:30, the timestamp 1:00:30 is inserted into the log. After the insertion, the log size is 2, not larger than the allowed count. Thus, the request is allowed.

- A new request arrives at 1:00:50, and the timestamp is inserted into the log. After the insertion, the log size is 3, larger than the allowed size 2. Therefore, this request is rejected even though the timestamp remains in the log.

- A new request arrives at 1:01:40. Requests in the range [1:00:40,1:01:40) are within the latest time frame, but requests sent before 1:00:40 are outdated. Two outdated timestamps, 1:00:01 and 1:00:30, are removed from the log. After the remove operation, the log size becomes 2; therefore, the request is accepted.

Pros:

- Rate limiting implemented by this algorithm is very accurate. In any rolling window, requests will not exceed the rate limit.

Cons:

- The algorithm consumes a lot of memory because even if a request is rejected, its timestamp might still be stored in memory.

Sliding window counter algorithm

The sliding window counter algorithm is a hybrid approach that combines the fixed window counter and sliding window log. The algorithm can be implemented by two different approaches. We will explain one implementation in this section and provide reference for the other implementation at the end of the section. Figure 4-11 illustrates how this algorithm works.

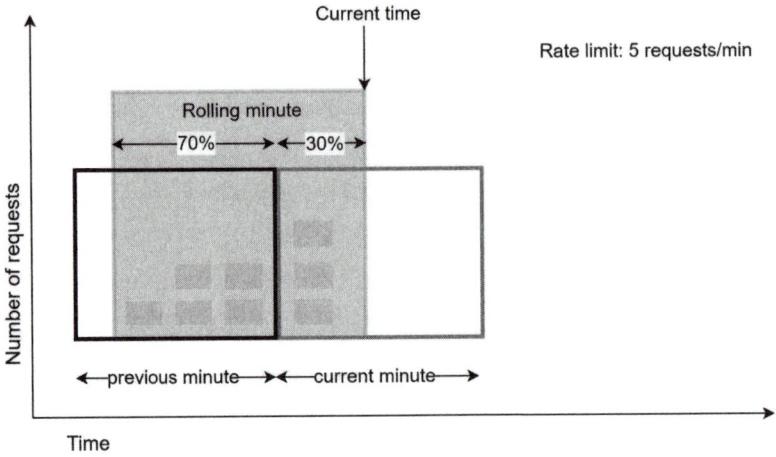

Figure 4-11

Assume the rate limiter allows a maximum of 7 requests per minute, and there are 5 requests in the previous minute and 3 in the current minute. For a new request that arrives at a 30% position in the current minute, the number of requests in the rolling window is calculated using the following formula:

- Requests in current window + requests in the previous window * overlap percentage of the rolling window and previous window

- Using this formula, we get 3 + 5 * 0.7% = 6.5 request. Depending on the use case, the number can either be rounded up or down. In our example, it is rounded down to 6.

Since the rate limiter allows a maximum of 7 requests per minute, the current request can go through. However, the limit will be reached after receiving one more request.

Due to the space limitation, we will not discuss the other implementation here. Interested readers should refer to the reference material [9]. This algorithm is not perfect. It has pros and cons.

Pros

- It smooths out spikes in traffic because the rate is based on the average rate of the previous window.
- Memory efficient.

Cons

- It only works for not-so-strict look back window. It is an approximation of the actual rate because it assumes requests in the previous window are evenly distributed. However, this problem may not be as bad as it seems. According to experiments done by Cloudflare [10], only 0.003% of requests are wrongly allowed or rate limited among 400 million requests.

High-level architecture

The basic idea of rate limiting algorithms is simple. At the high-level, we need a counter to keep track of how many requests are sent from the same user, IP address, etc. If the counter is larger than the limit, the request is disallowed.

Where shall we store counters? Using the database is not a good idea due to slowness of disk access. In-memory cache is chosen because it is fast and supports time-based expiration strategy. For instance, Redis [11] is a popular option to implement rate limiting. It is an in-memory store that offers two commands: INCR and EXPIRE.

- INCR: It increases the stored counter by 1.
- EXPIRE: It sets a timeout for the counter. If the timeout expires, the counter is automatically deleted.

Figure 4-12 shows the high-level architecture for rate limiting, and this works as follows:

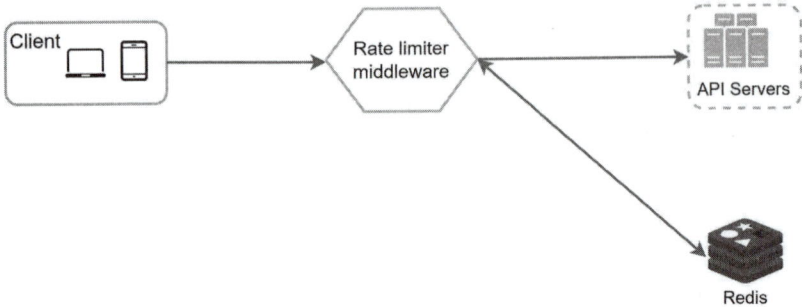

Figure 4-12

- The client sends a request to rate limiting middleware.

- Rate limiting middleware fetches the counter from the corresponding bucket in Redis and checks if the limit is reached or not.

    o   If the limit is reached, the request is rejected.

    o   If the limit is not reached, the request is sent to API servers. Meanwhile, the system increments the counter and saves it back to Redis.

Step 3 - Design deep dive

The high-level design in Figure 4-12 does not answer the following questions:

- How are rate limiting rules created? Where are the rules stored?

- How to handle requests that are rate limited?

In this section, we will first answer the questions regarding rate limiting rules and then go over the strategies to handle rate-limited requests. Finally, we will discuss rate limiting in distributed environment, a detailed design, performance optimization and monitoring.

Rate limiting rules

Lyft open-sourced their rate-limiting component [12]. We will peek inside of the component and look at some examples of rate limiting rules:

```
domain: messaging
descriptors:
  - key: message_type
    Value: marketing
    rate_limit:
      unit: day
      requests_per_unit: 5
```

In the above example, the system is configured to allow a maximum of 5 marketing messages per day. Here is another example:

```
domain: auth
descriptors:
  - key: auth_type
    Value: login
    rate_limit:
      unit: minute
      requests_per_unit: 5
```

This rule shows that clients are not allowed to login more than 5 times in 1 minute. Rules are generally written in configuration files and saved on disk.

Exceeding the rate limit

In case a request is rate limited, APIs return a HTTP response code 429 (too many requests) to the client. Depending on the use cases, we may enqueue the rate-limited requests to be processed later. For example, if some orders are rate limited due to system overload, we may keep those orders to be processed later.

Rate limiter headers

How does a client know whether it is being throttled? And how does a client know the number of allowed remaining requests before being throttled? The answer lies in HTTP response headers. The rate limiter returns the following HTTP headers to clients:

X-Ratelimit-Remaining: The remaining number of allowed requests within the window.

X-Ratelimit-Limit: It indicates how many calls the client can make per time window.

X-Ratelimit-Retry-After: The number of seconds to wait until you can make a request again without being throttled.

When a user has sent too many requests, a 429 too many requests error and *X-Ratelimit-Retry-After* header are returned to the client.

Detailed design

Figure 4-13 presents a detailed design of the system.

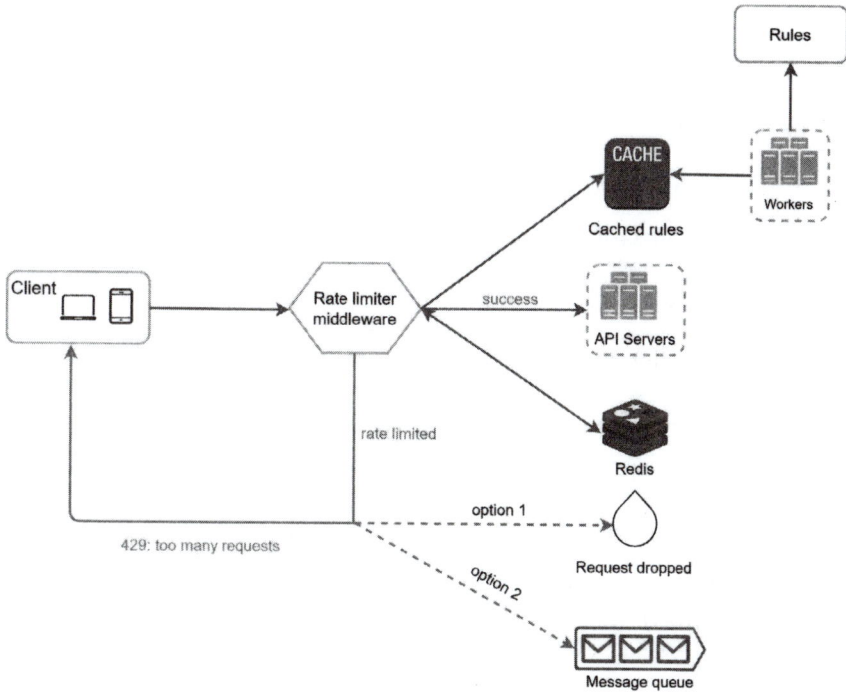

Figure 4-13

- Rules are stored on the disk. Workers frequently pull rules from the disk and store them in the cache.

- When a client sends a request to the server, the request is sent to the rate limiter middleware first.

- Rate limiter middleware loads rules from the cache. It fetches counters and last request timestamp from Redis cache. Based on the response, the rate limiter decides:

  o   if the request is not rate limited, it is forwarded to API servers.

  o   if the request is rate limited, the rate limiter returns 429 too many requests error to the client. In the meantime, the request is either dropped or forwarded to the queue.

Rate limiter in a distributed environment

Building a rate limiter that works in a single server environment is not difficult. However, scaling the system to support multiple servers and concurrent threads is a different story. There are two challenges:

- Race condition
- Synchronization issue

Race condition

As discussed earlier, rate limiter works as follows at the high-level:

- Read the *counter* value from Redis.
- Check if (*counter + 1*) exceeds the threshold.
- If not, increment the counter value by 1 in Redis.

Race conditions can happen in a highly concurrent environment as shown in Figure 4-14.

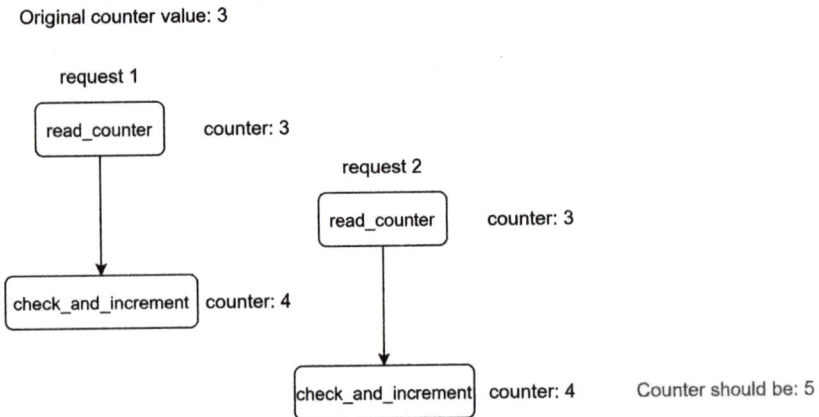

Figure 4-14

Assume the *counter* value in Redis is 3. If two requests concurrently read the *counter* value before either of them writes the value back, each will increment the *counter* by one and write it back without checking the other thread. Both requests (threads) believe they have the correct *counter* value 4. However, the correct *counter* value should be 5.

Locks are the most obvious solution for solving race condition. However, locks will significantly slow down the system. Two strategies are commonly used to solve the problem: Lua script [13] and sorted sets data structure in Redis [8]. For readers interested in these strategies, refer to the corresponding reference materials [8] [13].

Synchronization issue

Synchronization is another important factor to consider in a distributed environment. To support millions of users, one rate limiter server might not be enough to handle the traffic. When multiple rate limiter servers are used, synchronization is required. For example, on the left side of Figure 4-15, client 1 sends requests to rate limiter 1, and client 2 sends requests to rate limiter 2. As the web tier is stateless, clients can send requests to a different rate limiter as shown on the right side of Figure 4-15. If no synchronization happens, rate limiter 1 does not contain any data about client 2. Thus, the rate limiter cannot work properly.

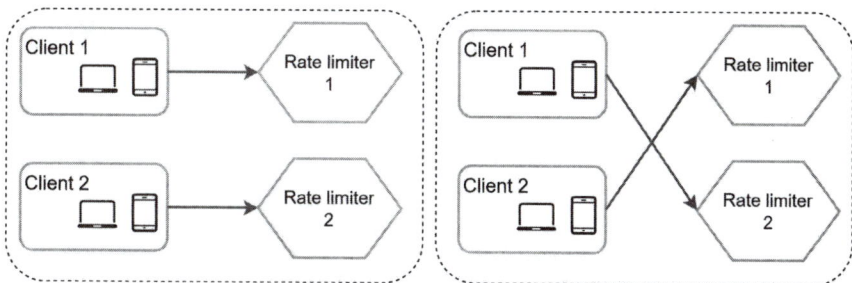

Figure 4-15

One possible solution is to use sticky sessions that allow a client to send traffic to the same rate limiter. This solution is not advisable because it is neither scalable nor flexible. A better approach is to use centralized data stores like Redis. The design is shown in Figure 4-16.

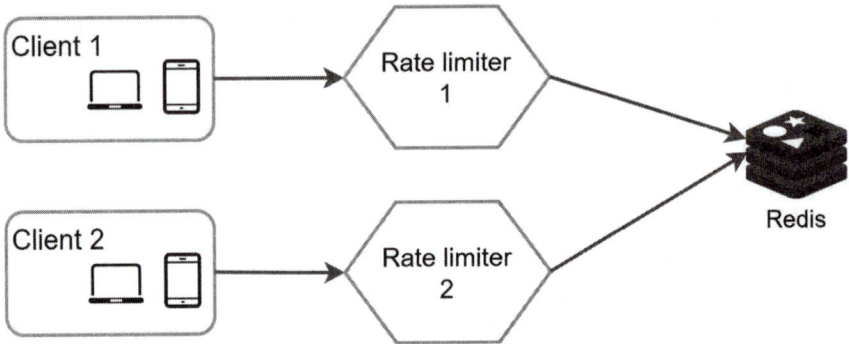

Figure 4-16

Performance optimization

Performance optimization is a common topic in system design interviews. We will cover two areas to improve.

First, multi-data center setup is crucial for a rate limiter because latency is high for users located far away from the data center. Most cloud service providers build many edge server locations around the world. For example, as of 5/20 2020, Cloudflare has 194 geographically distributed edge servers [14]. Traffic is automatically routed to the closest edge server to reduce latency.

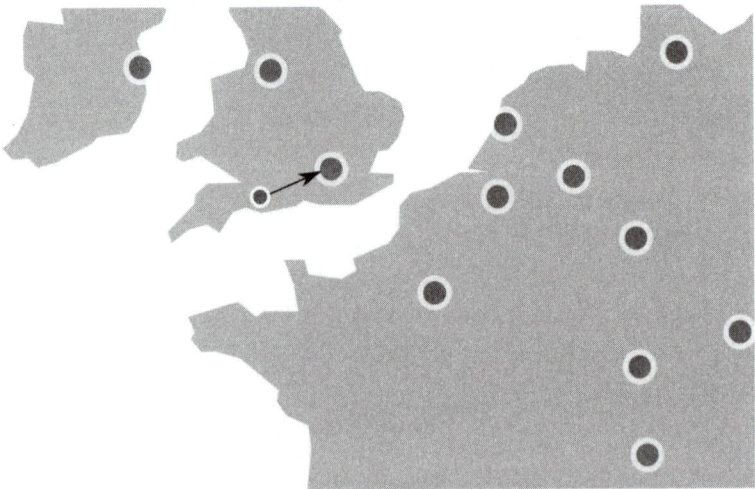

Figure 4-17 (Source: [10])

Second, synchronize data with an eventual consistency model. If you are unclear about the eventual consistency model, refer to the "Consistency" section in "Chapter 6: Design a Key-value Store."

Monitoring

After the rate limiter is put in place, it is important to gather analytics data to check whether the rate limiter is effective. Primarily, we want to make sure:

- The rate limiting algorithm is effective.
- The rate limiting rules are effective.

For example, if rate limiting rules are too strict, many valid requests are dropped. In this case, we want to relax the rules a little bit. In another example, we notice our rate limiter becomes ineffective when there is a sudden increase in traffic like flash sales. In this scenario, we may replace the algorithm to support burst traffic. Token bucket is a good fit here.

Step 4 - Wrap up

In this chapter, we discussed different algorithms of rate limiting and their pros/cons. Algorithms discussed include:

- Token bucket
- Leaking bucket
- Fixed window
- Sliding window log
- Sliding window counter

Then, we discussed the system architecture, rate limiter in a distributed environment, performance optimization and monitoring. Similar to any system design interview questions, there are additional talking points you can mention if time allows:

- Hard vs soft rate limiting.
  - o Hard: The number of requests cannot exceed the threshold.

- o   Soft: Requests can exceed the threshold for a short period.

- Rate limiting at different levels. In this chapter, we only talked about rate limiting at the application level (HTTP: layer 7). It is possible to apply rate limiting at other layers. For example, you can apply rate limiting by IP addresses using Iptables [15] (IP: layer 3). Note: The Open Systems Interconnection model (OSI model) has 7 layers [16]: Layer 1: Physical layer, Layer 2: Data link layer, Layer 3: Network layer, Layer 4: Transport layer, Layer 5: Session layer, Layer 6: Presentation layer, Layer 7: Application layer.

- Avoid being rate limited. Design your client with best practices:

  - o   Use client cache to avoid making frequent API calls.

  - o   Understand the limit and do not send too many requests in a short time frame.

  - o   Include code to catch exceptions or errors so your client can gracefully recover from exceptions.

  - o   Add sufficient back off time to retry logic.

Congratulations on getting this far! Now give yourself a pat on the back. Good job!

Reference materials

[1] Rate-limiting strategies and techniques:
https://cloud.google.com/solutions/rate-limiting-strategies-techniques

[2] Twitter rate limits:
https://developer.twitter.com/en/docs/basics/rate-limits

[3] Google docs usage limits:
https://developers.google.com/docs/api/limits

[4] IBM microservices: https://www.ibm.com/cloud/learn/microservices

[5] Throttle API requests for better throughput:
https://docs.aws.amazon.com/apigateway/latest/developerguide/
api-gateway-request-throttling.html

[6] Stripe rate limiters: https://stripe.com/blog/rate-limiters

[7] Shopify REST Admin API rate limits:
https://help.shopify.com/en/api/reference/rest-admin-api-rate-limits

[8] Better Rate Limiting With Redis Sorted Sets:
https://engineering.classdojo.com/blog/2015/02/06/rolling-rate-limiter/

[9] System Design — Rate limiter and Data modelling:
https://medium.com/@saisandeepmopuri/system-design-rate-limiter-
and-data-modelling-9304b0d18250

[10] How we built rate limiting capable of scaling to millions of domains:
https://blog.cloudflare.com/counting-things-a-lot-of-different-things/

[11] Redis website: https://redis.io/

[12] Lyft rate limiting: https://github.com/lyft/ratelimit

[13] Scaling your API with rate limiters:
https://gist.github.com/ptarjan/e38f45f2dfe601419ca3af937fff-
574d#request-rate-limiter

[14] What is edge computing:
https://www.cloudflare.com/learning/serverless/glossary/what-is-edge-computing/

[15] Rate Limit Requests with Iptables:
https://blog.programster.org/rate-limit-requests-with-iptables

[16] OSI model:
https://en.wikipedia.org/wiki/OSI_model#Layer_architecture

5

DESIGN CONSISTENT HASHING

To achieve horizontal scaling, it is important to distribute requests/data efficiently and evenly across servers. Consistent hashing is a commonly used technique to achieve this goal. But first, let us take an in-depth look at the problem.

The rehashing problem

If you have *n* cache servers, a common way to balance the load is to use the following hash method:

serverIndex = hash(key) % N, where *N* is the size of the server pool.

Let us use an example to illustrate how it works. As shown in Table 5-1, we have 4 servers and 8 string keys with their hashes.

key	hash	hash % 4
key0	18358617	1
key1	26143584	0
key2	18131146	2
key3	35863496	0
key4	34085809	1
key5	27581703	3
key6	38164978	2
key7	22530351	3

Table 5-1

To fetch the server where a key is stored, we perform the modular operation *f(key) % 4*. For instance, *hash(key0) % 4 = 1* means a client must contact server 1 to fetch the cached data. Figure 5-1 shows the distribution of keys based on Table 5-1.

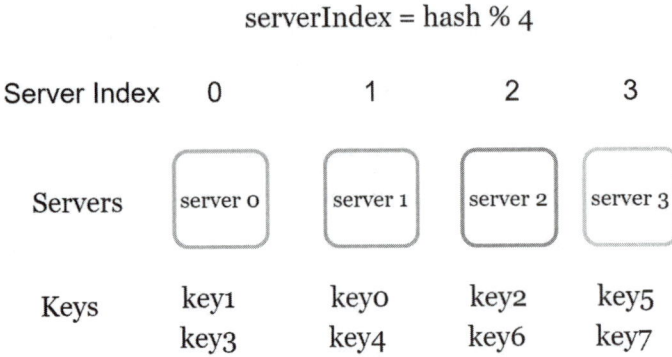

serverIndex = hash % 4

Server Index	0	1	2	3
Servers	server 0	server 1	server 2	server 3
Keys	key1 key3	key0 key4	key2 key6	key5 key7

Figure 5-1

This approach works well when the size of the server pool is fixed, and the data distribution is even. However, problems arise when new servers are added, or existing servers are removed. For example, if server 1 goes offline, the size of the server pool becomes 3. Using the same hash function, we get the same hash value for a key. But applying modular operation gives us different server indexes because the number of servers is reduced by 1. We get the results as shown in Table 5-2 by applying *hash % 3*:

key	hash	hash % 3
key0	18358617	0
key1	26143584	0
key2	18131146	1
key3	35863496	2
key4	34085809	1
key5	27581703	0
key6	38164978	1
key7	22530351	0

Table 5-2

Figure 5-2 shows the new distribution of keys based on Table 5-2.

$$serverIndex = hash \% 3$$

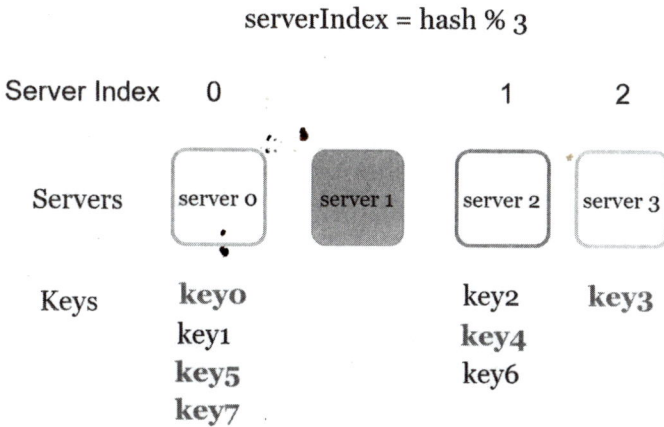

Figure 5-2

As shown in Figure 5-2, most keys are redistributed, not just the ones originally stored in the offline server (server 1). This means that when server 1 goes offline, most cache clients will connect to the wrong servers to fetch data. This causes a storm of cache misses. Consistent hashing is an effective technique to mitigate this problem.

Consistent hashing

Quoted from Wikipedia: "Consistent hashing is a special kind of hashing such that when a hash table is re-sized and consistent hashing is used, only k/n keys need to be remapped on average, where k is the number of keys, and n is the number of slots. In contrast, in most traditional hash tables, a change in the number of array slots causes nearly all keys to be remapped [1]".

Hash space and hash ring

Now we understand the definition of consistent hashing, let us find out how it works. Assume SHA-1 is used as the hash function f, and the output range of the hash function is: *x0, x1, x2, x3, …, xn*. In cryptography, SHA-1's hash space goes from 0 to $2^{160} - 1$. That means *x0* corresponds to 0, *xn* corresponds to $2^{160} - 1$, and all the other hash values in the middle fall between 0 and $2^{160} - 1$. Figure 5-3 shows the hash space.

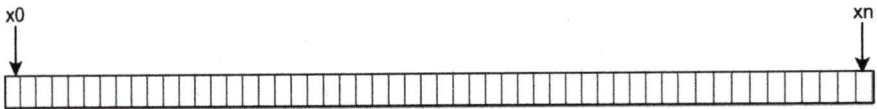

Figure 5-3

By connecting both ends, we get a hash ring as shown in Figure 5-4:

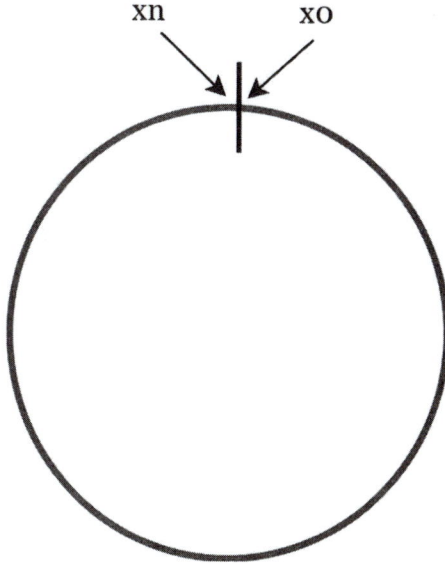

Figure 5-4

Hash servers

Using the same hash function f, we map servers based on server IP or name onto the ring. Figure 5-5 shows that 4 servers are mapped on the hash ring.

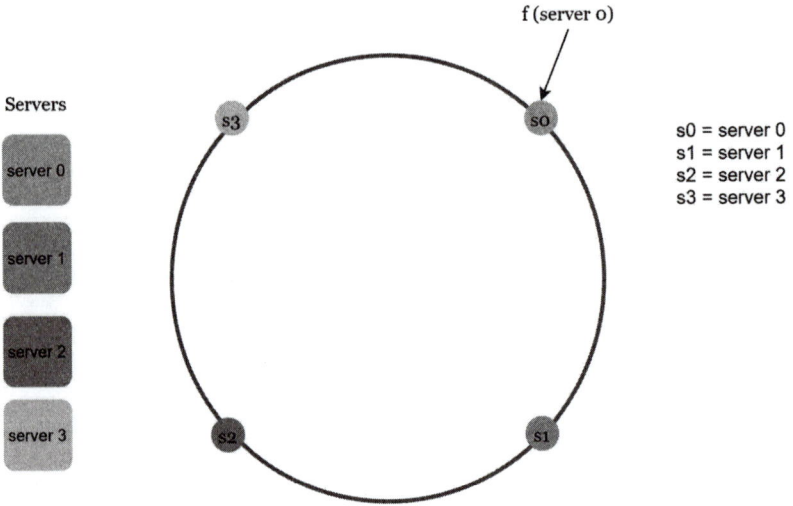

Figure 5-5

Hash keys

One thing worth mentioning is that hash function used here is different from the one in "the rehashing problem," and there is no modular operation. As shown in Figure 5-6, 4 cache keys (key0, key1, key2, and key3) are hashed onto the hash ring

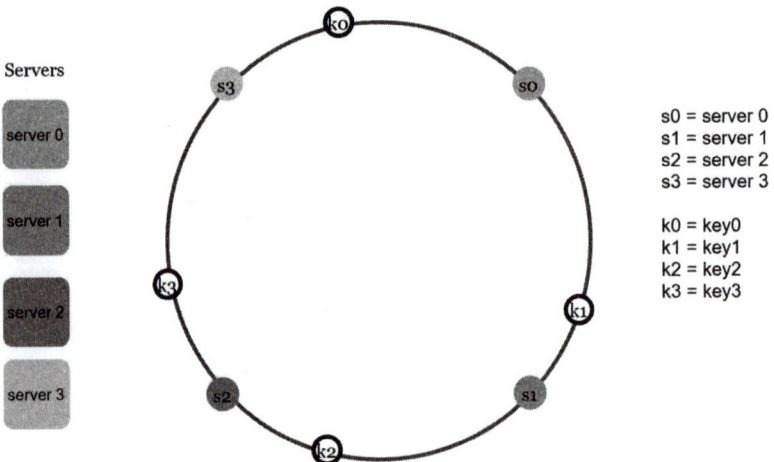

Figure 5-6

Server lookup

To determine which server a key is stored on, we go clockwise from the key position on the ring until a server is found. Figure 5-7 explains this process. Going clockwise, *key0* is stored on *server 0*; *key1* is stored on *server 1*; *key2* is stored on *server 2* and *key3* is stored on *server 3*.

Servers

server 0

server 1

server 2

server 3

s0 = server 0
s1 = server 1
s2 = server 2
s3 = server 3

k0 = key0
k1 = key1
k2 = key2
k3 = key3

Figure 5-7

Add a server

Using the logic described above, adding a new server will only require redistribution of a fraction of keys.

In Figure 5-8, after a new *server 4* is added, only *key0* needs to be redistributed. *k1*, *k2*, and *k3* remain on the same servers. Let us take a close look at the logic. Before *server 4* is added, *key0* is stored on *server 0*. Now, *key0* will be stored on *server 4* because *server 4* is the first server it encounters by going clockwise from *key0*'s position on the ring. The other keys are not redistributed based on consistent hashing algorithm.

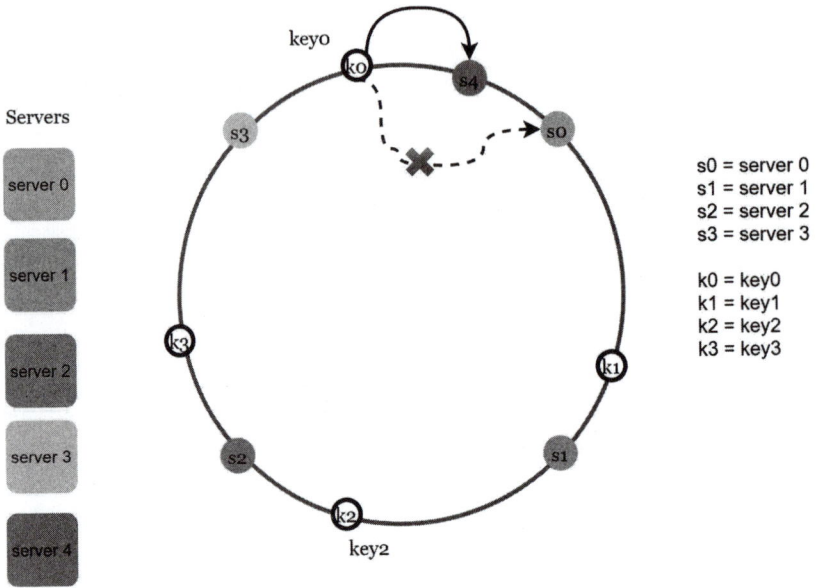

keyo

Servers

s3

s0

s0 = server 0
s1 = server 1
s2 = server 2
s3 = server 3

server 0

server 1

k0 = key0
k1 = key1
k2 = key2
k3 = key3

server 2

k3

k1

server 3

s2

s1

server 4

k2

key2

Figure 5-8

Remove a server

When a server is removed, only a small fraction of keys require redistribution with consistent hashing. In Figure 5-9, when *server 1* is removed, only *key1* must be remapped to *server 2*. The rest of the keys are unaffected.

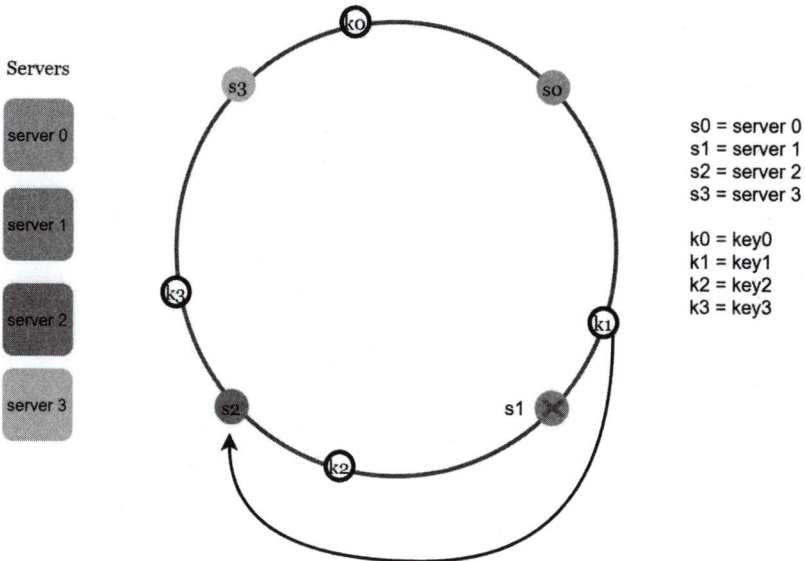

Servers

server 0

server 1

server 2

server 3

s0 = server 0
s1 = server 1
s2 = server 2
s3 = server 3

k0 = key0
k1 = key1
k2 = key2
k3 = key3

Figure 5-9

Two issues in the basic approach

The consistent hashing algorithm was introduced by Karger et al. at MIT [1]. The basic steps are:

- Map servers and keys on to the ring using a uniformly distributed hash function.

- To find out which server a key is mapped to, go clockwise from the key position until the first server on the ring is found.

Two problems are identified with this approach. First, it is impossible to keep the same size of partitions on the ring for all servers considering a server can be added or removed. A partition is the hash space between adjacent servers. It is possible that the size of the partitions on the ring assigned to each server is very small or fairly large. In Figure 5-10, if *s1* is removed, *s2's* partition (highlighted with the bidirectional arrows) is twice as large as *s0* and *s3's* partition.

Figure 5-10

Second, it is possible to have a non-uniform key distribution on the ring. For instance, if servers are mapped to positions listed in Figure 5-11, most of the keys are stored on *server 2*. However, *server 1* and *server 3* have no data.

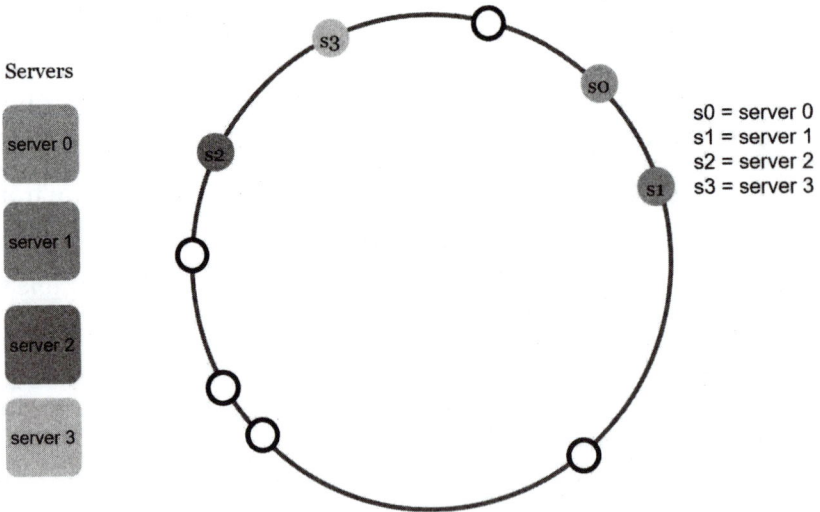

Figure 5-11

A technique called virtual nodes or replicas is used to solve these problems.

Virtual nodes

A virtual node refers to the real node, and each server is represented by multiple virtual nodes on the ring. In Figure 5-12, both *server 0* and *server 1* have 3 virtual nodes. The 3 is arbitrarily chosen; and in real-world systems, the number of virtual nodes is much larger. Instead of using *s0*, we have *s0_0, s0_1*, and s0_2 to represent *server 0* on the ring. Similarly, *s1_0, s1_1*, and *s1_2* represent server 1 on the ring. With virtual nodes, each server is responsible for multiple partitions. Partitions (edges) with label *s0* are managed by server 0. On the other hand, partitions with label *s1* are managed by *server 1*.

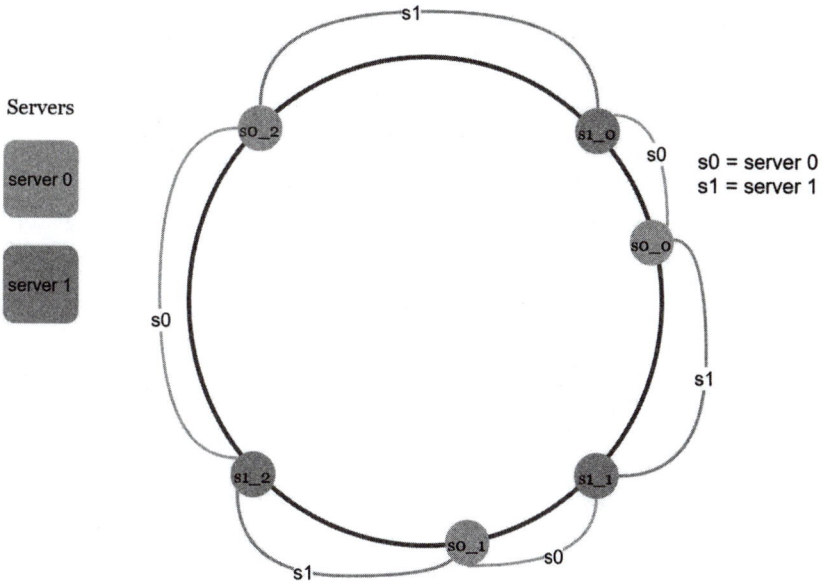

Figure 5-12

To find which server a key is stored on, we go clockwise from the key's location and find the first virtual node encountered on the ring. In Figure 5-13, to find out which server *k0* is stored on, we go clockwise from *k0*'s location and find virtual node *s1_1*, which refers to *server 1*.

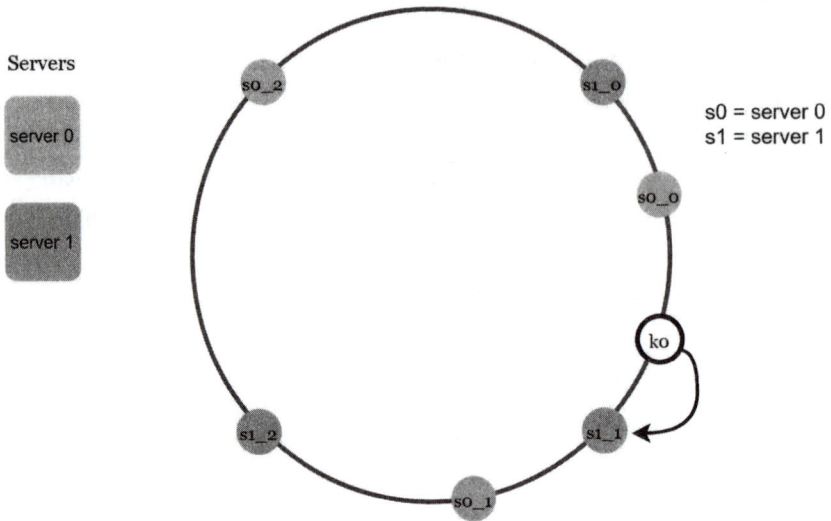

Figure 5-13

As the number of virtual nodes increases, the distribution of keys becomes more balanced. This is because the standard deviation gets smaller with more virtual nodes, leading to balanced data distribution. Standard deviation measures how data are spread out. The outcome of an experiment carried out by online research [2] shows that with one or two hundred virtual nodes, the standard deviation is between 5% (200 virtual nodes) and 10% (100 virtual nodes) of the mean. The standard deviation will be smaller when we increase the number of virtual nodes. However, more spaces are needed to store data about virtual nodes. This is a tradeoff, and we can tune the number of virtual nodes to fit our system requirements.

Find affected keys

When a server is added or removed, a fraction of data needs to be redistributed. How can we find the affected range to redistribute the keys?

In Figure 5-14, *server 4* is added onto the ring. The affected range starts from *s4* (newly added node) and moves anticlockwise around the ring until a server is found (*s3*). Thus, keys located between *s3* and *s4* need to be redistributed to *s4*.

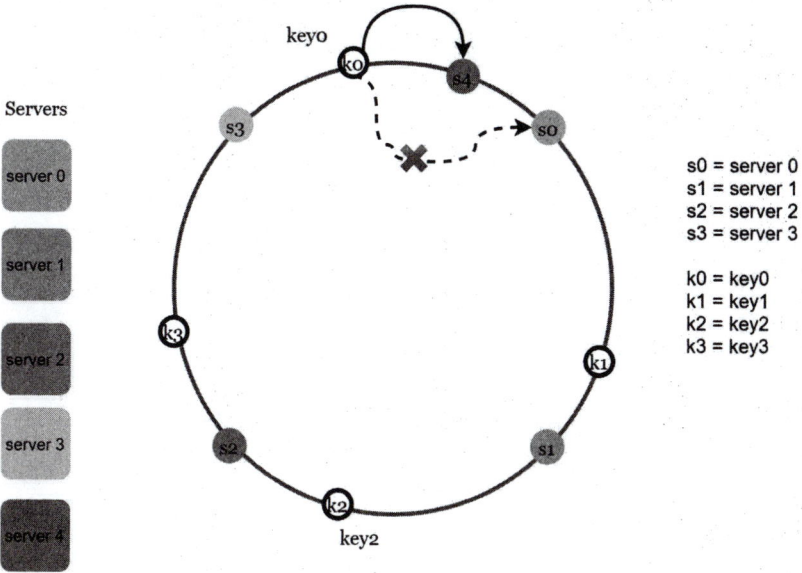

keyo

Servers

s0 = server 0
s1 = server 1
s2 = server 2
s3 = server 3

k0 = key0
k1 = key1
k2 = key2
k3 = key3

key2

Figure 5-14

When a server (*s1*) is removed as shown in Figure 5-15, the affected range starts from *s1* (removed node) and moves anticlockwise around the ring until a server is found (*s0*). Thus, keys located between *s0* and *s1* must be redistributed to *s2*.

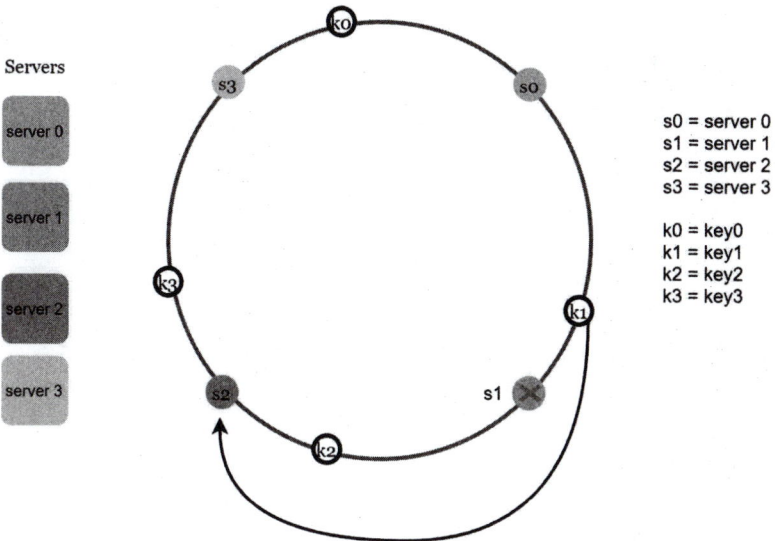

Servers

s0 = server 0
s1 = server 1
s2 = server 2
s3 = server 3

k0 = key0
k1 = key1
k2 = key2
k3 = key3

Figure 5-15

Wrap up

In this chapter, we had an in-depth discussion about consistent hashing, including why it is needed and how it works. The benefits of consistent hashing include:

- Minimized keys are redistributed when servers are added or removed.

- It is easy to scale horizontally because data are more evenly distributed.

- Mitigate hotspot key problem. Excessive access to a specific shard could cause server overload. Imagine data for Katy Perry, Justin Bieber, and Lady Gaga all end up on the same shard. Consistent hashing helps to mitigate the problem by distributing the data more evenly.

Consistent hashing is widely used in real-world systems, including some notable ones:

- Partitioning component of Amazon's Dynamo database [3]

- Data partitioning across the cluster in Apache Cassandra [4]

- Discord chat application [5]

- Akamai content delivery network [6]

- Maglev network load balancer [7]

Congratulations on getting this far! Now give yourself a pat on the back. Good job!

Reference materials

[1] Consistent hashing:
https://en.wikipedia.org/wiki/Consistent_hashing

[2] Consistent Hashing:
https://tom-e-white.com/2007/11/consistent-hashing.html

[3] Dynamo: Amazon's Highly Available Key-value Store:
https://www.allthingsdistributed.com/files/amazon-dynamo-sosp2007.pdf

[4] Cassandra - A Decentralized Structured Storage System:
http://www.cs.cornell.edu/Projects/ladis2009/papers/Lakshman-ladis2009.PDF

[5] How Discord Scaled Elixir to 5,000,000 Concurrent Users:
https://blog.discord.com/scaling-elixir-f9b8e1e7c29b

[6] CS168: The Modern Algorithmic Toolbox Lecture #1: Introduction
and Consistent Hashing: http://theory.stanford.edu/~tim/s16/l/l1.pdf

[7] Maglev: A Fast and Reliable Software Network Load Balancer:
https://static.googleusercontent.com/media/research.google.com/en//pubs/archive/44824.pdf

6

DESIGN A KEY-VALUE STORE

A key-value store, also referred to as a key-value database, is a non-relational database. Each unique identifier is stored as a key with its associated value. This data pairing is known as a "key-value" pair.

In a key-value pair, the key must be unique, and the value associated with the key can be accessed through the key. Keys can be plain text or hashed values. For performance reasons, a short key works better. What do keys look like? Here are a few examples:

- Plain text key: "last_logged_in_at"
- Hashed key: 253DDEC4

The value in a key-value pair can be strings, lists, objects, etc. The value is usually treated as an opaque object in key-value stores, such as Amazon dynamo [1], Memcached [2], Redis [3], etc.

Here is a data snippet in a key-value store:

key	value
145	john
147	bob
160	julia

Table 6-1

In this chapter, you are asked to design a key-value store that supports the following operations:

- put(key, value) // insert "value" associated with "key"

- get(key) // get "value" associated with "key"

Understand the problem and establish design scope

There is no perfect design. Each design achieves a specific balance regarding the tradeoffs of the read, write, and memory usage. Another tradeoff has to be made was between consistency and availability. In this chapter, we design a key-value store that comprises of the following characteristics:

- The size of a key-value pair is small: less than 10 KB.

- Ability to store big data.

- High availability: The system responds quickly, even during failures.

- High scalability: The system can be scaled to support large data set.

- Automatic scaling: The addition/deletion of servers should be automatic based on traffic.

- Tunable consistency.

- Low latency.

Single server key-value store

Developing a key-value store that resides in a single server is easy. An intuitive approach is to store key-value pairs in a hash table, which keeps everything in memory. Even though memory access is fast, fitting everything in memory may be impossible due to the space constraint. Two optimizations can be done to fit more data in a single server:

- Data compression

- Store only frequently used data in memory and the rest on disk

Even with these optimizations, a single server can reach its capacity very quickly. A distributed key-value store is required to support big data.

Distributed key-value store

A distributed key-value store is also called a distributed hash table, which distributes key-value pairs across many servers. When designing a distributed system, it is important to understand CAP (**C**onsistency, **A**vailability, **P**artition Tolerance) theorem.

CAP theorem

CAP theorem states it is impossible for a distributed system to simultaneously provide more than two of these three guarantees: consistency, availability, and partition tolerance. Let us establish a few definitions.

Consistency: consistency means all clients see the same data at the same time no matter which node they connect to.

Availability: availability means any client which requests data gets a response even if some of the nodes are down.

Partition Tolerance: a partition indicates a communication break between two nodes. Partition tolerance means the system continues to operate despite network partitions.

CAP theorem states that one of the three properties must be sacrificed to support 2 of the 3 properties as shown in Figure 6-1.

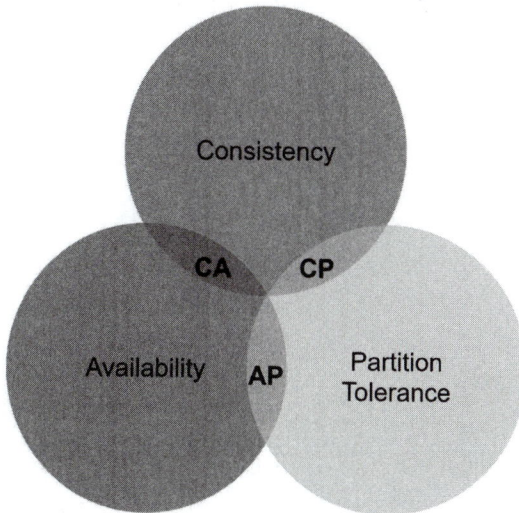

Figure 6-1

Nowadays, key-value stores are classified based on the two CAP charac-teristics they support:

CP (consistency and partition tolerance) systems: a CP key-value store supports consistency and partition tolerance while sacrificing availability.

AP (availability and partition tolerance) systems: an AP key-value store supports availability and partition tolerance while sacrificing consistency.

CA (consistency and availability) systems: a CA key-value store sup-ports consistency and availability while sacrificing partition tolerance. Since network failure is unavoidable, a distributed system must toler-ate network partition. Thus, a CA system cannot exist in real-world ap-plications.

What you read above is mostly the definition part. To make it easier to understand, let us take a look at some concrete examples. In distributed systems, data is usually replicated multiple times. Assume data are repli-cated on three replica nodes, *n1*, *n2* and *n3* as shown in Figure 6-2.

Ideal situation

In the ideal world, network partition never occurs. Data written to *n1* is automatically replicated to *n2* and *n3*. Both consistency and availability are achieved.

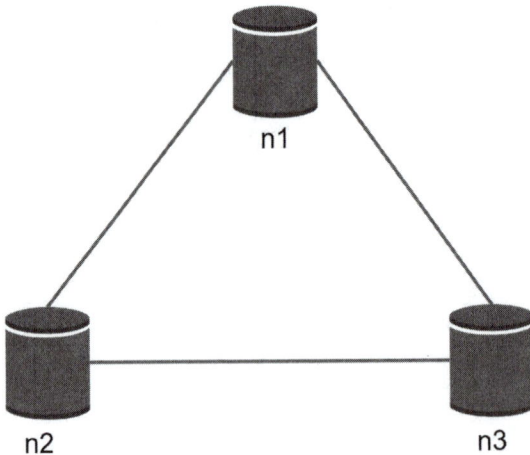

Figure 6-2

Real-world distributed systems

In a distributed system, partitions cannot be avoided, and when a partition occurs, we must choose between consistency and availability. In Figure 6-3, *n3* goes down and cannot communicate with *n1* and *n2*. If clients write data to *n1* or *n2*, data cannot be propagated to n3. If data is written to *n3* but not propagated to *n1* and *n2* yet, *n1* and *n2* would have stale data.

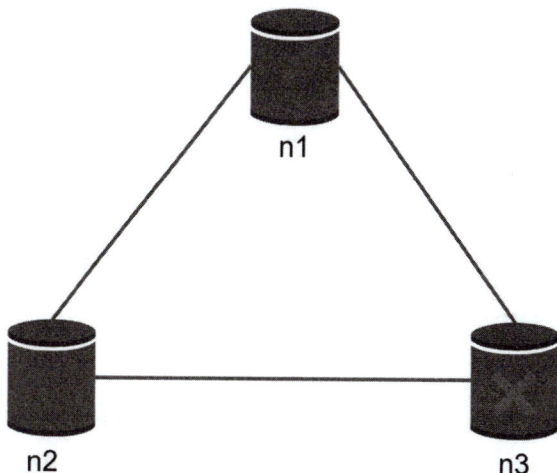

Figure 6-3

If we choose consistency over availability (CP system), we must block all write operations to *n1* and *n2* to avoid data inconsistency among these three servers, which makes the system unavailable. Bank systems usually have extremely high consistent requirements. For example, it is crucial for a bank system to display the most up-to-date balance info. If inconsistency occurs due to a network partition, the bank system returns an error before the inconsistency is resolved.

However, if we choose availability over consistency (AP system), the system keeps accepting reads, even though it might return stale data. For writes, *n1* and *n2* will keep accepting writes, and data will be synced to *n3* when the network partition is resolved.

Choosing the right CAP guarantees that fit your use case is an important step in building a distributed key-value store. You can discuss this with your interviewer and design the system accordingly.

System components

In this section, we will discuss the following core components and techniques used to build a key-value store:

- Data partition
- Data replication
- Consistency
- Inconsistency resolution
- Handling failures
- System architecture diagram
- Write path
- Read path

The content below is largely based on three popular key-value store systems: Dynamo [4], Cassandra [5], and BigTable [6].

Data partition

For large applications, it is infeasible to fit the complete data set in a single server. The simplest way to accomplish this is to split the data into smaller partitions and store them in multiple servers. There are two challenges while partitioning the data:

- Distribute data across multiple servers evenly.
- Minimize data movement when nodes are added or removed.

Consistent hashing discussed in Chapter 5 is a great technique to solve these problems. Let us revisit how consistent hashing works at a high-level.

- First, servers are placed on a hash ring. In Figure 6-4, eight servers, represented by *s0, s1, ..., s7*, are placed on the hash ring.
- Next, a key is hashed onto the same ring, and it is stored on the

first server encountered while moving in the clockwise direction. For instance, *key0* is stored in *s1* using this logic.

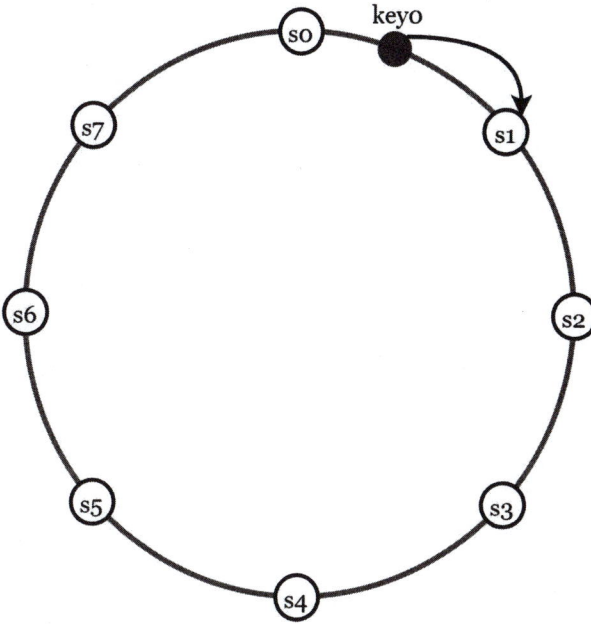

Figure 6-4

Using consistent hashing to partition data has the following advantages:

Automatic scaling: servers could be added and removed automatically depending on the load.

Heterogeneity: the number of virtual nodes for a server is proportional to the server capacity. For example, servers with higher capacity are assigned with more virtual nodes.

Data replication

To achieve high availability and reliability, data must be replicated asynchronously over N servers, where N is a configurable parameter. These N servers are chosen using the following logic: after a key is mapped to a position on the hash ring, walk clockwise from that position and choose the first *N* servers on the ring to store data copies. In Figure 6-5 (*N = 3*), *key0* is replicated at *s1, s2,* and *s3*.

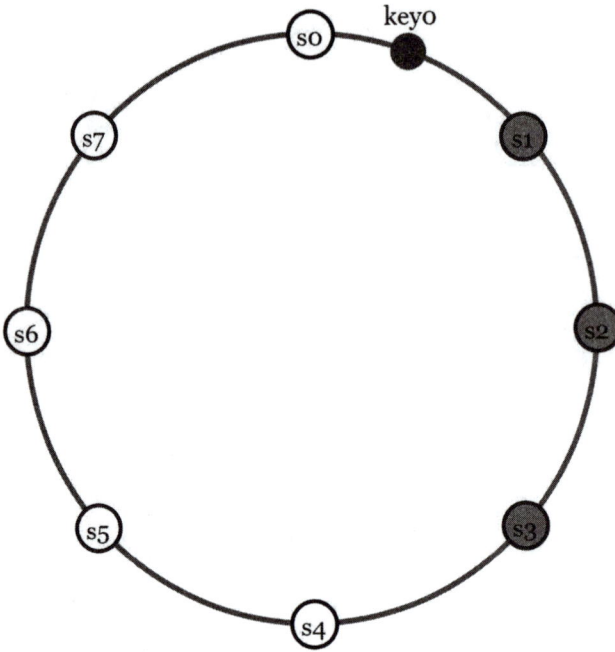

Figure 6-5

With virtual nodes, the first N nodes on the ring may be owned by fewer than N physical servers. To avoid this issue, we only choose unique servers while performing the clockwise walk logic.

Nodes in the same data center often fail at the same time due to power outages, network issues, natural disasters, etc. For better reliability, replicas are placed in distinct data centers, and data centers are connected through high-speed networks.

Consistency

Since data is replicated at multiple nodes, it must be synchronized across replicas. Quorum consensus can guarantee consistency for both read and write operations. Let us establish a few definitions first.

N = The number of replicas

W = A write quorum of size W. For a write operation to be considered as successful, write operation must be acknowledged from W replicas.

R = A read quorum of size R. For a read operation to be considered as successful, read operation must wait for responses from at least R replicas.

Consider the following example shown in Figure 6-6 with $N = 3$.

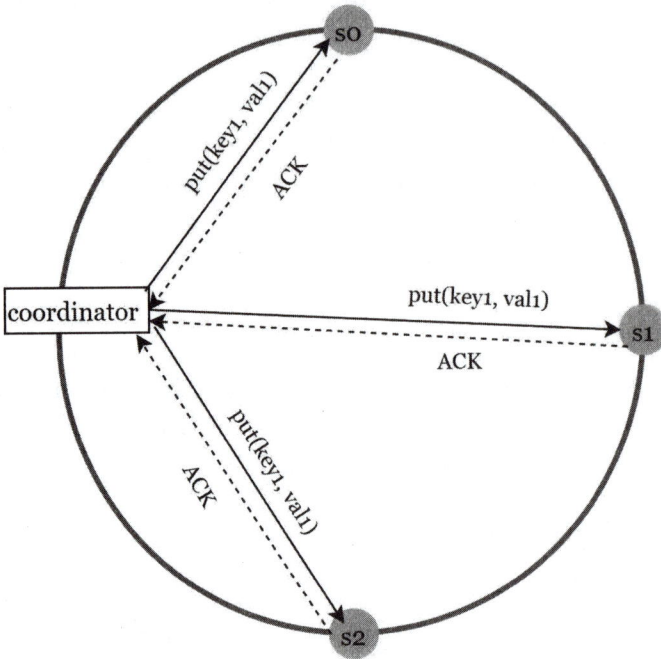

Figure 6-6 (ACK = acknowledgement)

$W = 1$ does not mean data is written on one server. For instance, with the configuration in Figure 6-6, data is replicated at $s0$, $s1$, and $s2$. $W = 1$ means that the coordinator must receive at least one acknowledgment before the write operation is considered as successful. For instance, if we get an acknowledgment from $s1$, we no longer need to wait for acknowledgements from $s0$ and $s2$. A coordinator acts as a proxy between the client and the nodes.

The configuration of W, R and N is a typical tradeoff between latency and consistency. If $W = 1$ or $R = 1$, an operation is returned quickly because a coordinator only needs to wait for a response from any of the replicas. If W or $R > 1$, the system offers better consistency; however, the query will be slower because the coordinator must wait for the response from the slowest replica.

If $W + R > N$, strong consistency is guaranteed because there must be at least one overlapping node that has the latest data to ensure consistency.

How to configure N, W, and R to fit our use cases? Here are some of the possible setups:

If $R = 1$ and $W = N$, the system is optimized for a fast read.

If $W = 1$ and $R = N$, the system is optimized for fast write.

If $W + R > N$, strong consistency is guaranteed (Usually $N = 3$, $W = R = 2$).

If $W + R <= N$, strong consistency is not guaranteed.

Depending on the requirement, we can tune the values of W, R, N to achieve the desired level of consistency.

Consistency models

Consistency model is other important factor to consider when designing a key-value store. A consistency model defines the degree of data consistency, and a wide spectrum of possible consistency models exist:

- Strong consistency: any read operation returns a value corresponding to the result of the most updated write data item. A client never sees out-of-date data.

- Weak consistency: subsequent read operations may not see the most updated value.

- Eventual consistency: this is a specific form of weak consistency. Given enough time, all updates are propagated, and all replicas are consistent.

Strong consistency is usually achieved by forcing a replica not to accept new reads/writes until every replica has agreed on current write. This approach is not ideal for highly available systems because it could block new operations. Dynamo and Cassandra adopt eventual consistency, which is our recommended consistency model for our key-value store. From concurrent writes, eventual consistency allows inconsistent values to enter the system and force the client to read the values to reconcile. The next section explains how reconciliation works with versioning.

Inconsistency resolution: versioning

Replication gives high availability but causes inconsistencies among replicas. Versioning and vector clocks are used to solve inconsistency problems. Versioning means treating each data modification as a new immutable version of data. Before we talk about versioning, let us use an example to explain how inconsistency happens:

As shown in Figure 6-7, both replica nodes *n1* and *n2* have the same value. Let us call this value the original *value*. *Server 1* and *server 2* get the same value for *get("name")* operation.

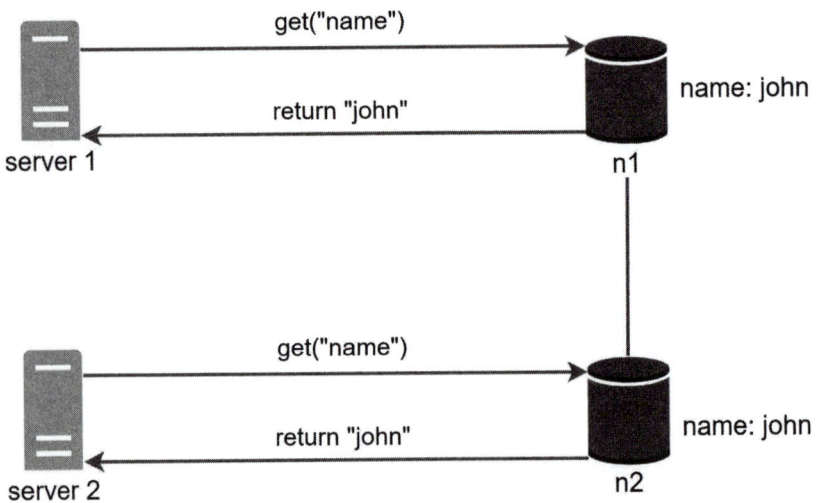

Figure 6-7

Next, *server 1* changes the name to "johnSanFrancisco", and *server 2* changes the name to "johnNewYork" as shown in Figure 6-8. These two changes are performed simultaneously. Now, we have conflicting values, called versions *v1* and *v2*.

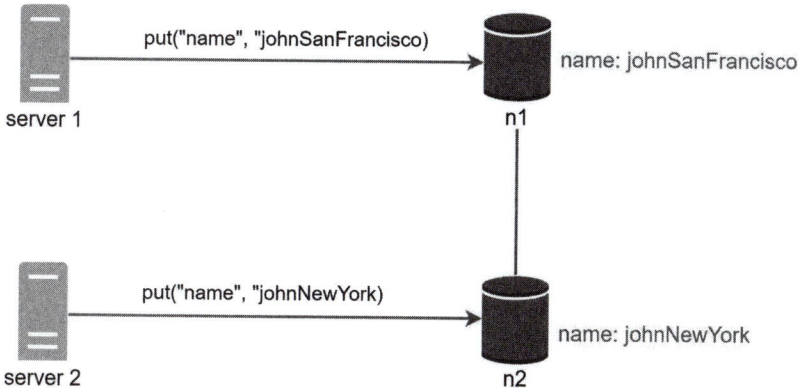

Figure 6-8

In this example, the original value could be ignored because the modifications were based on it. However, there is no clear way to resolve the conflict of the last two versions. To resolve this issue, we need a versioning system that can detect conflicts and reconcile conflicts. A vector clock is a common technique to solve this problem. Let us examine how vector clocks work.

A vector clock is a *[server, version]* pair associated with a data item. It can be used to check if one version precedes, succeeds, or in conflict with others.

Assume a vector clock is represented by $D([S1, v1], [S2, v2], ..., [Sn, vn])$, where D is a data item, $v1$ is a version counter, and $s1$ is a server number, etc. If data item D is written to server Si, the system must perform one of the following tasks.

- Increment vi if $[Si, vi]$ exists.

- Otherwise, create a new entry $[Si, 1]$.

The above abstract logic is explained with a concrete example as shown in Figure 6-9.

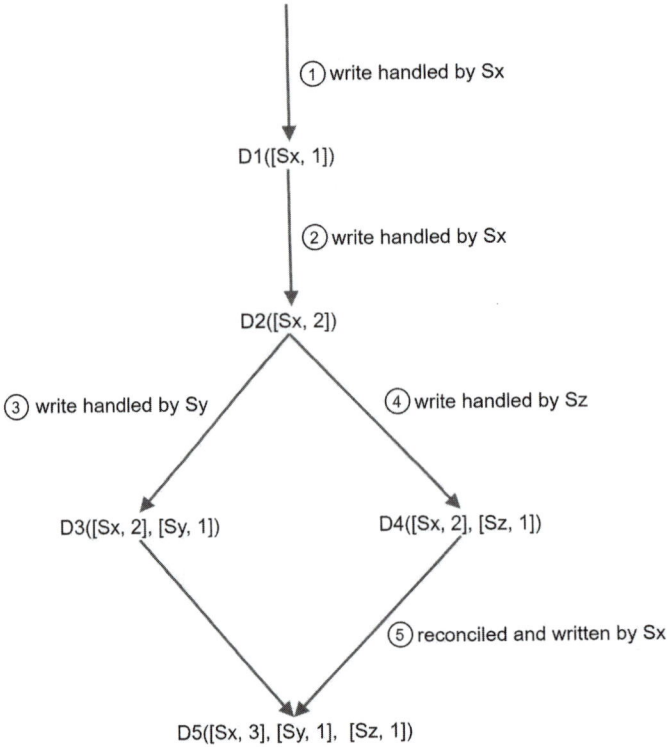

Figure 6-9

1. A client writes a data item *D1* to the system, and the write is handled by server *Sx*, which now has the vector clock *D1[(Sx, 1)]*.

2. Another client reads the latest *D1*, updates it to *D2*, and writes it back. *D2* descends from *D1* so it overwrites *D1*. Assume the write is handled by the same server *Sx*, which now has vector clock *D2([Sx, 2])*.

3. Another client reads the latest *D2*, updates it to *D3*, and writes it back. Assume the write is handled by server *Sy*, which now has vector clock *D3([Sx, 2], [Sy, 1]))*.

4. Another client reads the latest *D2*, updates it to *D4*, and writes it back. Assume the write is handled by server *Sz*, which now has *D4([Sx, 2], [Sz, 1]))*.

5. When another client reads *D3* and *D4*, it discovers a conflict, which is caused by data item *D2* being modified by both *Sy* and

Sz. The conflict is resolved by the client and updated data is sent to the server. Assume the write is handled by *Sx*, which now has $D5([Sx, 3], [Sy, 1], [Sz, 1])$. We will explain how to detect conflict shortly.

Using vector clocks, it is easy to tell that a version X is an ancestor (i.e. no conflict) of version Y if the version counters for each participant in the vector clock of Y is greater than or equal to the ones in version X. For example, the vector clock $D([s0, 1], [s1, 1])]$ is an ancestor of $D([s0, 1], [s1, 2])$. Therefore, no conflict is recorded.

Similarly, you can tell that a version X is a sibling (i.e., a conflict exists) of Y if there is any participant in Y's vector clock who has a counter that is less than its corresponding counter in X. For example, the following two vector clocks indicate there is a conflict: $D([s0, 1], [s1, 2])$ and $D([s0, 2], [s1, 1])$.

Even though vector clocks can resolve conflicts, there are two notable downsides. First, vector clocks add complexity to the client because it needs to implement conflict resolution logic.

Second, the *[server: version]* pairs in the vector clock could grow rapidly. To fix this problem, we set a threshold for the length, and if it exceeds the limit, the oldest pairs are removed. This can lead to inefficiencies in reconciliation because the descendant relationship cannot be determined accurately. However, based on Dynamo paper [4], Amazon has not yet encountered this problem in production; therefore, it is probably an acceptable solution for most companies.

Handling failures

As with any large system at scale, failures are not only inevitable but common. Handling failure scenarios is very important. In this section, we first introduce techniques to detect failures. Then, we go over common failure resolution strategies.

Failure detection

In a distributed system, it is insufficient to believe that a server is down because another server says so. Usually, it requires at least two independent sources of information to mark a server down.

As shown in Figure 6-10, all-to-all multicasting is a straightforward solution. However, this is inefficient when many servers are in the system.

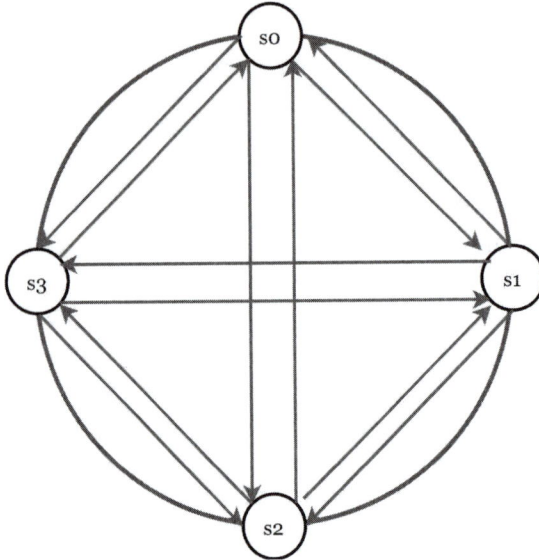

Figure 6-10

A better solution is to use decentralized failure detection methods like gossip protocol. Gossip protocol works as follows:

- Each node maintains a node membership list, which contains member IDs and heartbeat counters.

- Each node periodically increments its heartbeat counter.

- Each node periodically sends heartbeats to a set of random nodes, which in turn propagate to another set of nodes.

- Once nodes receive heartbeats, membership list is updated to the latest info.

- If the heartbeat has not increased for more than predefined periods, the member is considered as offline.

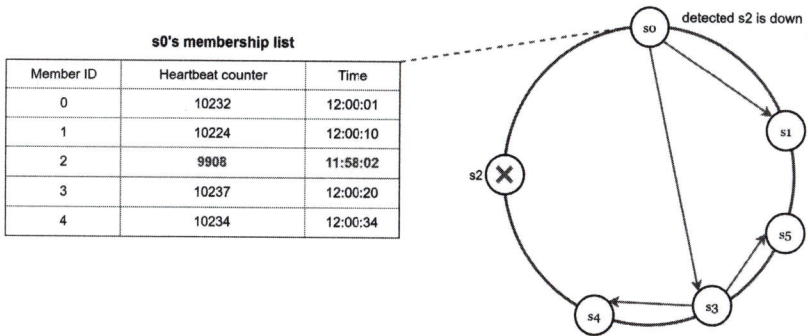

s0's membership list

Member ID	Heartbeat counter	Time
0	10232	12:00:01
1	10224	12:00:10
2	9908	11:58:02
3	10237	12:00:20
4	10234	12:00:34

Figure 6-11

As shown in Figure 6-11:

- Node *s0* maintains a node membership list shown on the left side.

- Node *s0* notices that node s2's (member ID = 2) heartbeat counter has not increased for a long time.

- Node *s0* sends heartbeats that include *s2*'s info to a set of random nodes. Once other nodes confirm that *s2*'s heartbeat counter has not been updated for a long time, node *s2* is marked down, and this information is propagated to other nodes.

Handling temporary failures

After failures have been detected through the gossip protocol, the system needs to deploy certain mechanisms to ensure availability. In the strict quorum approach, read and write operations could be blocked as illustrated in the quorum consensus section.

A technique called "sloppy quorum" [4] is used to improve availability. Instead of enforcing the quorum requirement, the system chooses the first *W* healthy servers for writes and first *R* healthy servers for reads on the hash ring. Offline servers are ignored.

If a server is unavailable due to network or server failures, another server will process requests temporarily. When the down server is up, changes will be pushed back to achieve data consistency. This process is called hinted handoff. Since *s2* is unavailable in Figure 6-12, reads and writes

will be handled by *s3* temporarily. When *s2* comes back online, *s3* will
hand the data back to *s2*.

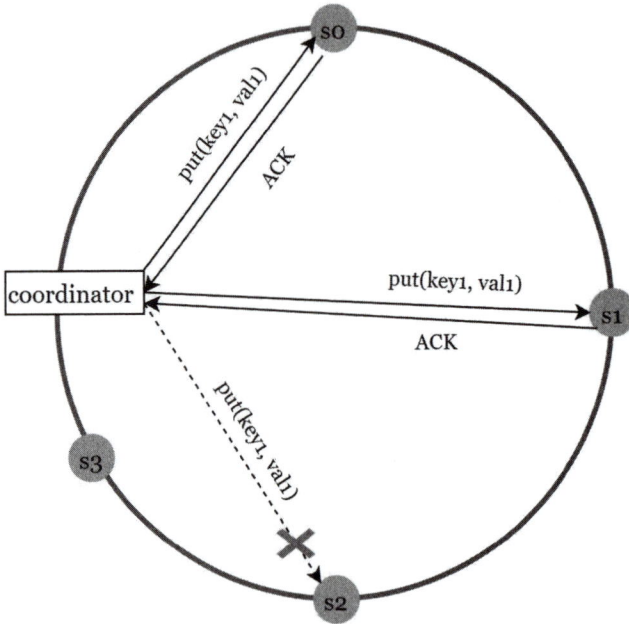

Figure 6-12

Handling permanent failures

Hinted handoff is used to handle temporary failures. What if a replica
is permanently unavailable? To handle such a situation, we implement
an anti-entropy protocol to keep replicas in sync. Anti-entropy involves
comparing each piece of data on replicas and updating each replica to
the newest version. A Merkle tree is used for inconsistency detection and
minimizing the amount of data transferred.

Quoted from Wikipedia [7]: "A hash tree or Merkle tree is a tree in which
every non-leaf node is labeled with the hash of the labels or values (in
case of leaves) of its child nodes. Hash trees allow efficient and secure
verification of the contents of large data structures".

Assuming key space is from 1 to 12, the following steps show how to
build a Merkle tree. Highlighted boxes indicate inconsistency.

Step 1: Divide key space into buckets (4 in our example) as shown in Figure 6-13. A bucket is used as the root level node to maintain a limited depth of the tree.

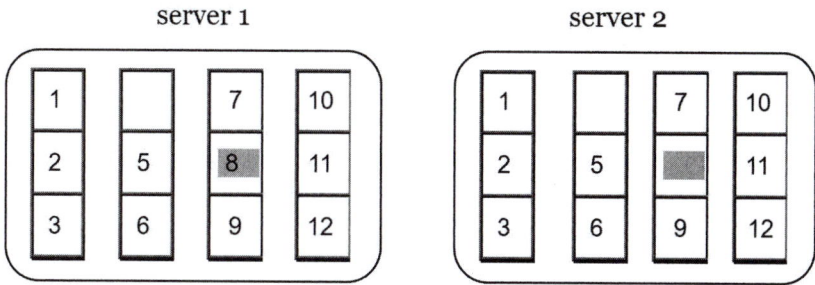

Figure 6-13

Step 2: Once the buckets are created, hash each key in a bucket using a uniform hashing method (Figure 6-14).

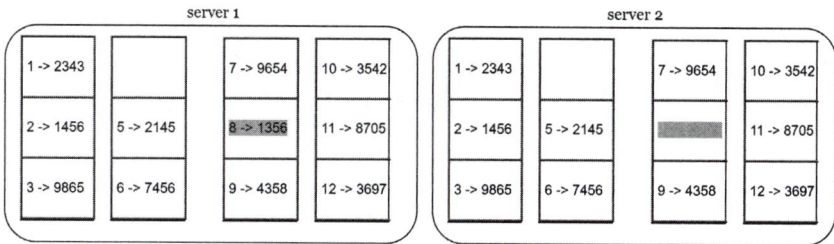

Figure 6-14

Step 3: Create a single hash node per bucket (Figure 6-15).

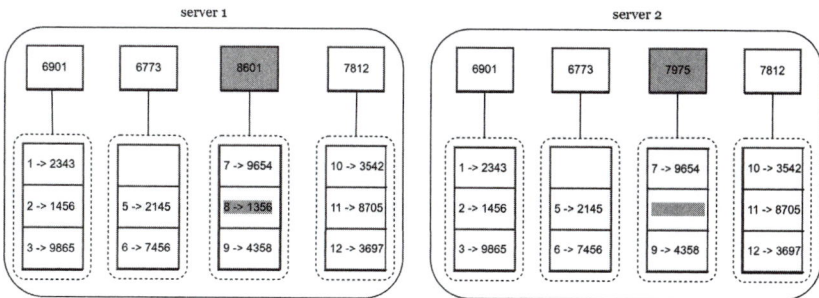

Figure 6-15

Step 4: Build the tree upwards till root by calculating hashes of children (Figure 6-16).

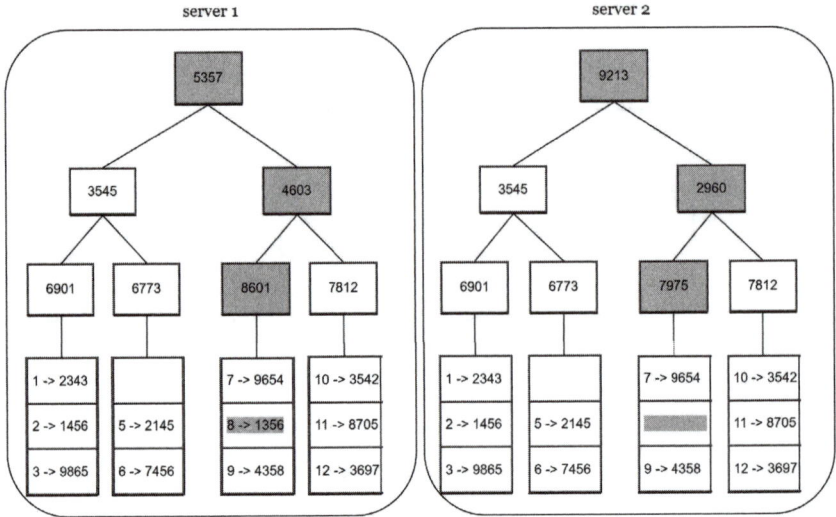

Figure 6-16

To compare two Merkle trees, start by comparing the root hashes. If root hashes match, both servers have the same data. If root hashes disagree, then the left child hashes are compared followed by right child hashes. You can traverse the tree to find which buckets are not synchronized and synchronize those buckets only.

Using Merkle trees, the amount of data needed to be synchronized is proportional to the differences between the two replicas, and not the amount of data they contain. In real-world systems, the bucket size is quite big. For instance, a possible configuration is one million buckets per one billion keys, so each bucket only contains 1000 keys.

Handling data center outage

Data center outage could happen due to power outage, network outage, natural disaster, etc. To build a system capable of handling data center outage, it is important to replicate data across multiple data centers. Even if a data center is completely offline, users can still access data through the other data centers.

System architecture diagram

Now that we have discussed different technical considerations in designing a key-value store, we can shift our focus on the architecture diagram, shown in Figure 6-17.

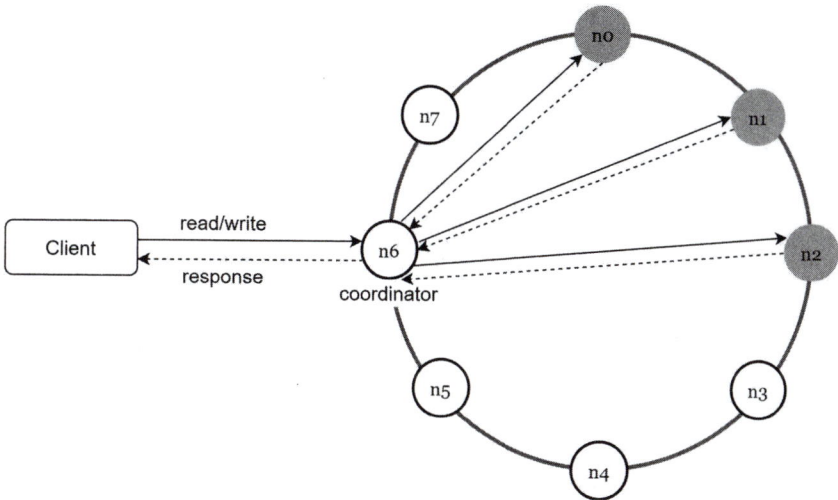

Figure 6-17

Main features of the architecture are listed as follows:

- Clients communicate with the key-value store through simple APIs: *get(key)* and *put(key, value)*.

- A coordinator is a node that acts as a proxy between the client and the key-value store.

- Nodes are distributed on a ring using consistent hashing.

- The system is completely decentralized so adding and moving nodes can be automatic.

- Data is replicated at multiple nodes.

- There is no single point of failure as every node has the same set of responsibilities.

As the design is decentralized, each node performs many tasks as presented in Figure 6-18.

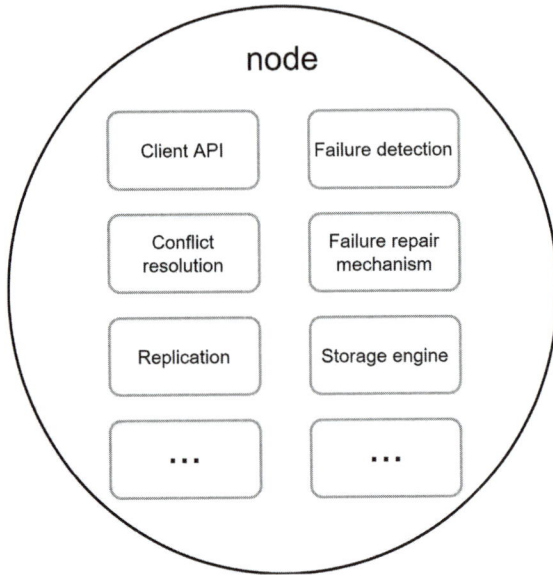

Figure 6-18

Write path

Figure 6-19 explains what happens after a write request is directed to a specific node. Please note the proposed designs for write/read paths are primary based on the architecture of Cassandra [8].

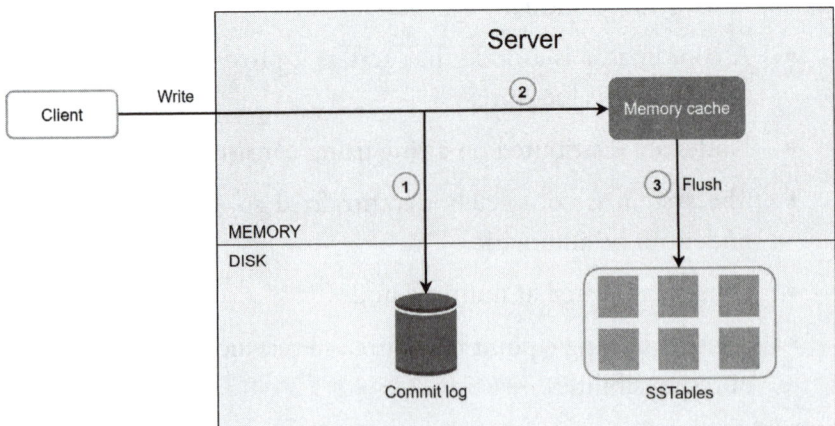

Figure 6-19

1. The write request is persisted on a commit log file.

2. Data is saved in the memory cache.

3. When the memory cache is full or reaches a predefined threshold, data is flushed to SSTable [9] on disk. Note: A sorted-string table (SSTable) is a sorted list of <key, value> pairs. For readers interested in learning more about SStable, refer to the reference material [9].

Read path

After a read request is directed to a specific node, it first checks if data is in the memory cache. If so, the data is returned to the client as shown in Figure 6-20.

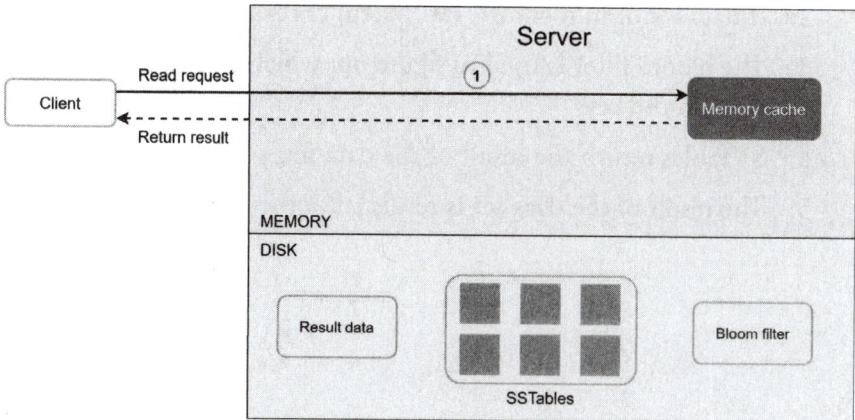

Figure 6-20

If the data is not in memory, it will be retrieved from the disk instead. We need an efficient way to find out which SSTable contains the key. Bloom filter [10] is commonly used to solve this problem.

The read path is shown in Figure 6-21 when data is not in memory.

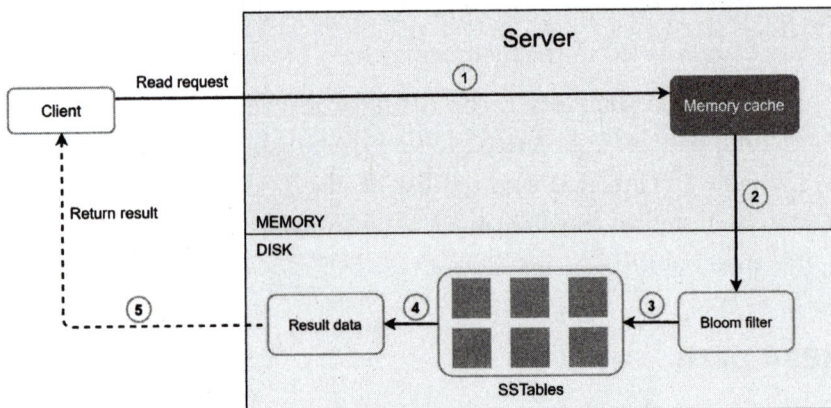

Figure 6-21

1. The system first checks if data is in memory. If not, go to step 2.

2. If data is not in memory, the system checks the bloom filter.

3. The bloom filter is used to figure out which SSTables might contain the key.

4. SSTables return the result of the data set.

5. The result of the data set is returned to the client.

Summary

This chapter covers many concepts and techniques. To refresh your memory, the following table summarizes features and corresponding techniques used for a distributed key-value store.

Goal/Problems	Technique
Ability to store big data	Use consistent hashing to spread load across servers
High availability reads	Data replication Multi-datacenter setup
Highly available writes	Versioning and conflict resolution with vector clocks
Dataset partition	Consistent Hashing
Incremental scalability	Consistent Hashing
Heterogeneity	Consistent Hashing
Tunable consistency	Quorum consensus
Handling temporary failures	Sloppy quorum and hinted handoff
Handling permanent failures	Merkle tree
Handling data center outage	Cross-datacenter replication

Table 6-2

Reference materials

[1] Amazon DynamoDB: https://aws.amazon.com/dynamodb/

[2] memcached: https://memcached.org/

[3] Redis: https://redis.io/

[4] Dynamo: Amazon's Highly Available Key-value Store: https://www.allthingsdistributed.com/files/amazon-dynamo-sosp2007.pdf

[5] Cassandra: https://cassandra.apache.org/

[6] Bigtable: A Distributed Storage System for Structured Data: https://static.googleusercontent.com/media/research.google.com/en//archive/bigtable-osdi06.pdf

[7] Merkle tree: https://en.wikipedia.org/wiki/Merkle_tree

[8] Cassandra architecture: https://cassandra.apache.org/doc/latest/architecture/

[9] SStable: https://www.igvita.com/2012/02/06/sstable-and-log-structured-storage-leveldb/

[10] Bloom filter https://en.wikipedia.org/wiki/Bloom_filter

7

DESIGN A UNIQUE ID GENERATOR IN DISTRIBUTED SYSTEMS

In this chapter, you are asked to design a unique ID generator in distributed systems. Your first thought might be to use a primary key with the *auto_increment* attribute in a traditional database. However, *auto_increment* does not work in a distributed environment because a single database server is not large enough and generating unique IDs across multiple databases with minimal delay is challenging.

Here are a few examples of unique IDs:

```
+-----------------------+
|  user_id              |
+-----------------------+
|   1227238262110117894 |
+-----------------------+
|   1241107244890099715 |
+-----------------------+
|   1243643959492173824 |
+-----------------------+
|   1247686501489692673 |
+-----------------------+
|   1567981766075453440 |
+-----------------------+
```

Figure 7-1

Step 1 - Understand the problem and establish design scope

Asking clarification questions is the first step to tackle any system design interview question. Here is an example of candidate-interviewer interaction:

Candidate: What are the characteristics of unique IDs?
Interviewer: IDs must be unique and sortable.

Candidate: For each new record, does ID increment by 1?
Interviewer: The ID increments by time but not necessarily only increments by 1. IDs created in the evening are larger than those created in the morning on the same day.

Candidate: Do IDs only contain numerical values?
Interviewer: Yes, that is correct.

Candidate: What is the ID length requirement?
Interviewer: IDs should fit into 64-bit.

Candidate: What is the scale of the system?
Interviewer: The system should be able to generate 10,000 IDs per second.

Above are some of the sample questions that you can ask your interviewer. It is important to understand the requirements and clarify ambiguities. For this interview question, the requirements are listed as follows:

- IDs must be unique.
- IDs are numerical values only.
- IDs fit into 64-bit.
- IDs are ordered by date.
- Ability to generate over 10,000 unique IDs per second.

Step 2 - Propose high-level design and get buy-in

Multiple options can be used to generate unique IDs in distributed systems. The options we considered are:

- Multi-master replication

- Universally unique identifier (UUID)
- Ticket server
- Twitter snowflake approach

Let us look at each of them, how they work, and the pros/cons of each option.

Multi-master replication

As shown in Figure 7-2, the first approach is multi-master replication.

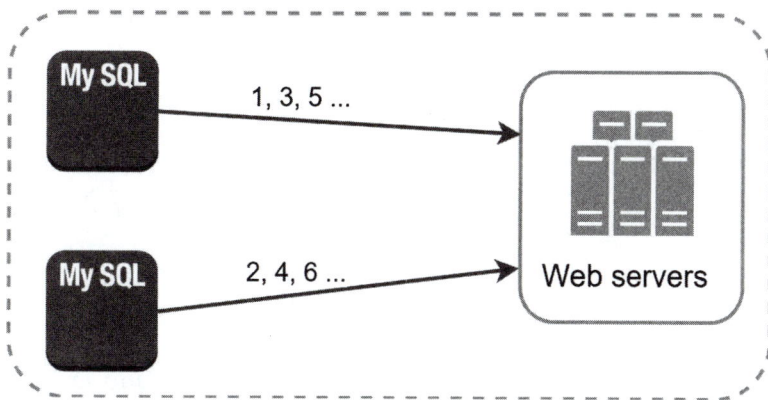

Figure 7-2

This approach uses the databases' *auto_increment* feature. Instead of increasing the next ID by 1, we increase it by k, where k is the number of database servers in use. As illustrated in Figure 7-2, next ID to be generated is equal to the previous ID in the same server plus 2. This solves some scalability issues because IDs can scale with the number of database servers. However, this strategy has some major drawbacks:

- Hard to scale with multiple data centers
- IDs do not go up with time across multiple servers.
- It does not scale well when a server is added or removed.

UUID

A UUID is another easy way to obtain unique IDs. UUID is a 128-bit number used to identify information in computer systems. UUID has a very low probability of getting collusion. Quoted from Wikipedia, "after generating 1 billion UUIDs every second for approximately 100 years would the probability of creating a single duplicate reach 50%" [1].

Here is an example of UUID: *09c93e62-50b4-468d-bf8a-c07e1040bfb2*. UUIDs can be generated independently without coordination between servers. Figure 7-3 presents the UUIDs design.

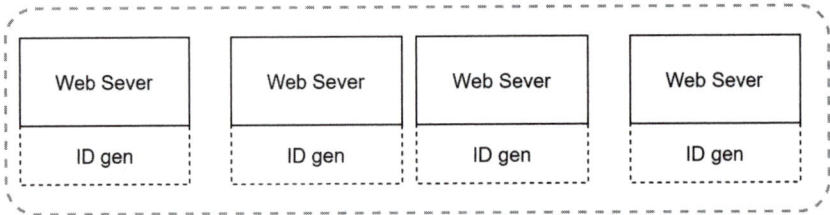

Figure 7-3

In this design, each web server contains an ID generator, and a web server is responsible for generating IDs independently.

Pros:

- Generating UUID is simple. No coordination between servers is needed so there will not be any synchronization issues.

- The system is easy to scale because each web server is responsible for generating IDs they consume. ID generator can easily scale with web servers.

Cons:

- IDs are 128 bits long, but our requirement is 64 bits.

- IDs do not go up with time.

- IDs could be non-numeric.

Ticket Server

Ticket servers are another interesting way to generate unique IDs. Flicker developed ticket servers to generate distributed primary keys [2]. It is worth mentioning how the system works.

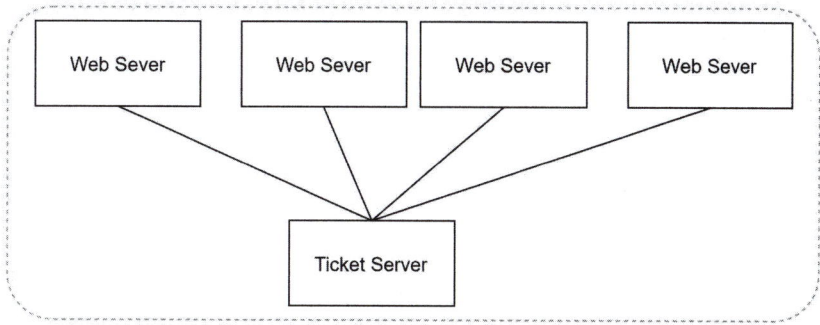

Figure 7-4

The idea is to use a centralized *auto_increment* feature in a single database server (Ticket Server). To learn more about this, refer to flicker's engineering blog article [2].

Pros:

- Numeric IDs.

- It is easy to implement, and it works for small to medium-scale applications.

Cons:

- Single point of failure. Single ticket server means if the ticket server goes down, all systems that depend on it will face issues. To avoid a single point of failure, we can set up multiple ticket servers. However, this will introduce new challenges such as data synchronization.

Twitter snowflake approach

Approaches mentioned above give us some ideas about how different ID generation systems work. However, none of them meet our specific requirements; thus, we need another approach. Twitter's unique ID generation system called "snowflake" [3] is inspiring and can satisfy our requirements.

Divide and conquer is our friend. Instead of generating an ID directly, we divide an ID into different sections. Figure 7-5 shows the layout of a 64-bit ID.

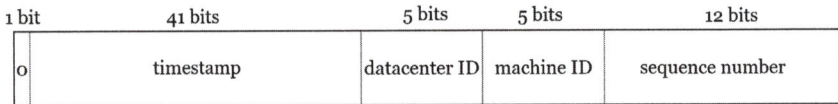

1 bit	41 bits	5 bits	5 bits	12 bits
0	timestamp	datacenter ID	machine ID	sequence number

Figure 7-5

Each section is explained below.

- Sign bit: 1 bit. It will always be 0. This is reserved for future uses. It can potentially be used to distinguish between signed and un-signed numbers.

- Timestamp: 41 bits. Milliseconds since the epoch or custom ep-och. We use Twitter snowflake default epoch 1288834974657, equivalent to Nov 04, 2010, 01:42:54 UTC.

- Datacenter ID: 5 bits, which gives us $2 \wedge 5 = 32$ datacenters.

- Machine ID: 5 bits, which gives us $2 \wedge 5 = 32$ machines per datacenter.

- Sequence number: 12 bits. For every ID generated on that ma-chine/process, the sequence number is incremented by 1. The number is reset to 0 every millisecond.

Step 3 - Design deep dive

In the high-level design, we discussed various options to design a unique ID generator in distributed systems. We settle on an approach that is based on the Twitter snowflake ID generator. Let us dive deep into the design. To refresh our memory, the design diagram is relisted below.

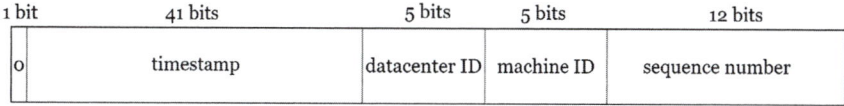

1 bit	41 bits	5 bits	5 bits	12 bits
0	timestamp	datacenter ID	machine ID	sequence number

Figure 7-6

Datacenter IDs and machine IDs are chosen at the startup time, generally fixed once the system is up running. Any changes in datacenter IDs and machine IDs require careful review since an accidental change in those values can lead to ID conflicts. Timestamp and sequence numbers are generated when the ID generator is running.

Timestamp

The most important 41 bits make up the timestamp section. As timestamps grow with time, IDs are sortable by time. Figure 7-7 shows an example of how binary representation is converted to UTC. You can also convert UTC back to binary representation using a similar method.

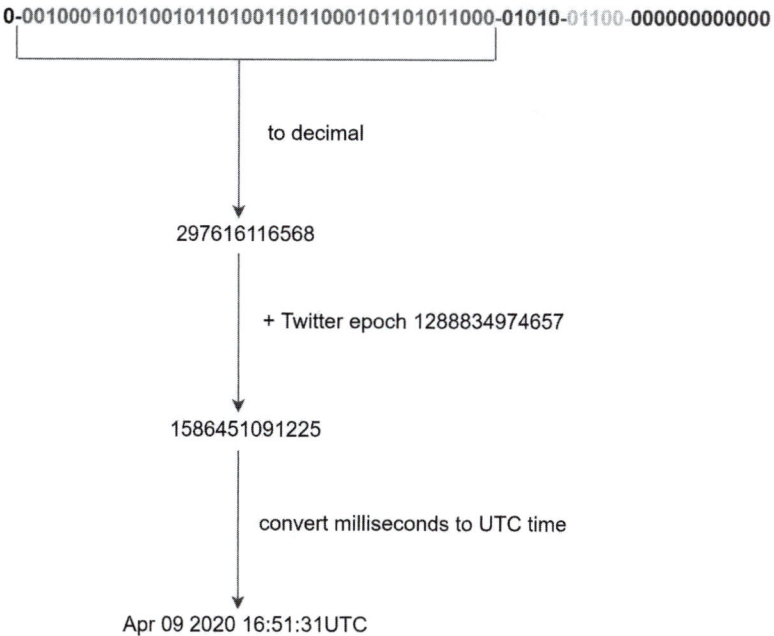

0-00100010101001011010011011000101101011000-01010-01100-000000000000

to decimal

297616116568

+ Twitter epoch 1288834974657

1586451091225

convert milliseconds to UTC time

Apr 09 2020 16:51:31UTC

Figure 7-7

The maximum timestamp that can be represented in 41 bits is

$2 \wedge 41 - 1 = 2199023255551$ milliseconds (ms), which gives us: ~ 69 years = *2199023255551 ms / 1000 / 365 days / 24 hours/ 3600 seconds*. This means the ID generator will work for 69 years and having a custom epoch time close to today's date delays the overflow time. After 69 years, we will need a new epoch time or adopt other techniques to migrate IDs.

Sequence number

Sequence number is 12 bits, which give us $2 \wedge 12 = 4096$ combinations. This field is 0 unless more than one ID is generated in a millisecond on the same server. In theory, a machine can support a maximum of 4096 new IDs per millisecond.

Step 4 - Wrap up

In this chapter, we discussed different approaches to design a unique ID generator: multi-master replication, UUID, ticket server, and Twitter snowflake-like unique ID generator. We settle on snowflake as it supports all our use cases and is scalable in a distributed environment.

If there is extra time at the end of the interview, here are a few additional talking points:

- Clock synchronization. In our design, we assume ID generation servers have the same clock. This assumption might not be true when a server is running on multiple cores. The same challenge exists in multi-machine scenarios. Solutions to clock synchronization are out of the scope of this book; however, it is important to understand the problem exists. Network Time Protocol is the most popular solution to this problem. For interested readers, refer to the reference material [4].

- Section length tuning. For example, fewer sequence numbers but more timestamp bits are effective for low concurrency and long-term applications.

- High availability. Since an ID generator is a mission-critical system, it must be highly available.

Congratulations on getting this far! Now give yourself a pat on the back. Good job!

Reference materials

[1] Universally unique identifier:
https://en.wikipedia.org/wiki/Universally_unique_identifier

[2] Ticket Servers: Distributed Unique Primary Keys on the Cheap:
https://code.flickr.net/2010/02/08/ticket-servers-distributed-unique-primary-keys-on-the-cheap/

[3] Announcing Snowflake:
https://blog.twitter.com/engineering/en_us/a/2010/announcing-snowflake.html

[4] Network time protocol:
https://en.wikipedia.org/wiki/Network_Time_Protocol

8

DESIGN A URL SHORTENER

In this chapter, we will tackle an interesting and classic system design interview question: designing a URL shortening service like tinyurl.

Step 1 - Understand the problem and establish design scope

System design interview questions are intentionally left open-ended. To design a well-crafted system, it is critical to ask clarification questions.

Candidate: Can you give an example of how a URL shortener work?
Interviewer: Assume URL https://www.systeminterview.com/ q=chatsystem&c=loggedin&v=v3&l=long is the original URL. Your service creates an alias with shorter length: https://tinyurl.com/ y7ke-ocwj. If you click the alias, it redirects you to the original URL.

Candidate: What is the traffic volume?
Interviewer: 100 million URLs are generated per day.

Candidate: How long is the shortened URL?
Interviewer: As short as possible.

Candidate: What characters are allowed in the shortened URL?
Interviewer: Shortened URL can be a combination of numbers (0-9) and characters (a-z, A-Z).

Candidate: Can shortened URLs be deleted or updated?
Interviewer: For simplicity, let us assume shortened URLs cannot be deleted or updated.

Here are the basic use cases:

1. URL shortening: given a long URL => return a much shorter URL

2. URL redirecting: given a shorter URL => redirect to the original URL

3. High availability, scalability, and fault tolerance considerations

Back of the envelope estimation

- Write operation: 100 million URLs are generated per day.

- Write operation per second: 100 million / 24 /3600 = 1160

- Read operation: Assuming ratio of read operation to write operation is 10:1, read operation per second: 1160 * 10 = 11,600

- Assuming the URL shortener service will run for 10 years, this means we must support 100 million * 365 * 10 = 365 billion records.

- Assume average URL length is 100.

- Storage requirement over 10 years: 365 billion * 100 bytes = 36.5 TB

It is important for you to walk through the assumptions and calculations with your interviewer so that both of you are on the same page.

Step 2 - Propose high-level design and get buy-in

In this section, we discuss the API endpoints, URL redirecting, and URL shortening flows.

API Endpoints

API endpoints facilitate the communication between clients and servers. We will design the APIs REST-style. If you are unfamiliar with restful API, you can consult external materials, such as the one in the reference material [1]. A URL shortener primary needs two API endpoints.

1.URL shortening. To create a new short URL, a client sends a POST request, which contains one parameter: the original long URL. The API looks like this:

POST api/v1/data/shorten

- request parameter: {longUrl: longURLString}
- return shortURL

2. URL redirecting. To redirect a short URL to the corresponding long URL, a client sends a GET request. The API looks like this:

GET api/v1/shortUrl

- Return longURL for HTTP redirection

URL redirecting

Figure 8-1 shows what happens when you enter a tinyurl onto the browser. Once the server receives a tinyurl request, it changes the short URL to the long URL with 301 redirect.

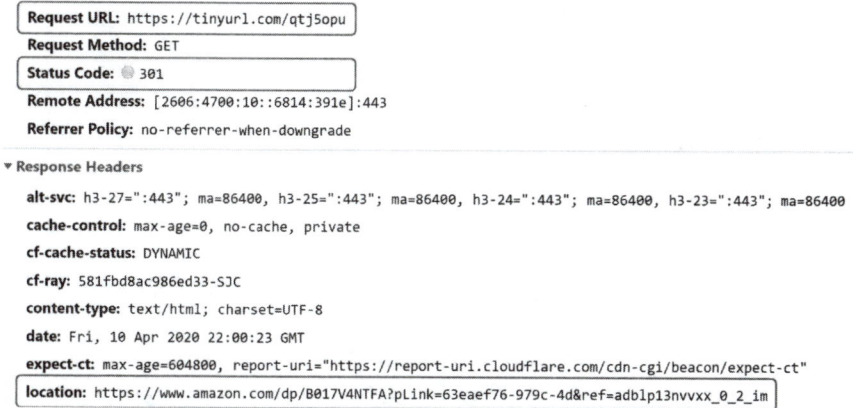

```
Request URL: https://tinyurl.com/qtj5opu
Request Method: GET
Status Code: ● 301
Remote Address: [2606:4700:10::6814:391e]:443
Referrer Policy: no-referrer-when-downgrade
```

▼ **Response Headers**

alt-svc: h3-27=":443"; ma=86400, h3-25=":443"; ma=86400, h3-24=":443"; ma=86400, h3-23=":443"; ma=86400
cache-control: max-age=0, no-cache, private
cf-cache-status: DYNAMIC
cf-ray: 581fbd8ac986ed33-SJC
content-type: text/html; charset=UTF-8
date: Fri, 10 Apr 2020 22:00:23 GMT
expect-ct: max-age=604800, report-uri="https://report-uri.cloudflare.com/cdn-cgi/beacon/expect-ct"
```
location: https://www.amazon.com/dp/B017V4NTFA?pLink=63eaef76-979c-4d&ref=adblp13nvvxx_0_2_im
```

Figure 8-1

The detailed communication between clients and servers is shown in Figure 8-2.

short URL: https://tinyurl.com/qtj5opu

long URL: https://www.amazon.com/dp/B017V4NTFA?pLink=63eaef76-979c-4d&
                 ref=adblp13nvvxx_0_2_im

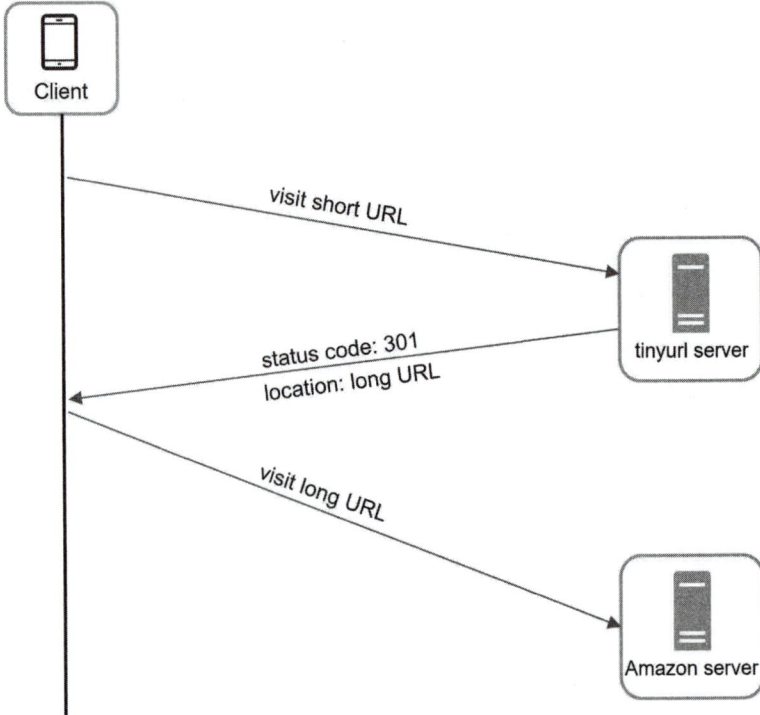

Figure 8-2

One thing worth discussing here is 301 redirect vs 302 redirect.

301 redirect. A 301 redirect shows that the requested URL is "permanently" moved to the long URL. Since it is permanently redirected, the browser caches the response, and subsequent requests for the same URL will not be sent to the URL shortening service. Instead, requests are redirected to the long URL server directly.

302 redirect. A 302 redirect means that the URL is "temporarily" moved to the long URL, meaning that subsequent requests for the same URL will be sent to the URL shortening service first. Then, they are redirected to the long URL server.

Each redirection method has its pros and cons. If the priority is to reduce the server load, using 301 redirect makes sense as only the first request of the same URL is sent to URL shortening servers. However, if analytics is important, 302 redirect is a better choice as it can track click rate and source of the click more easily.

The most intuitive way to implement URL redirecting is to use hash tables. Assuming the hash table stores *<shortURL, longURL>* pairs, URL redirecting can be implemented by the following:

- Get longURL: longURL = hashTable.get(shortURL)
- Once you get the longURL, perform the URL redirect.

URL shortening

Let us assume the short URL looks like this: www.tinyurl.com/{**hashValue**}. To support the URL shortening use case, we must find a hash function *fx* that maps a long URL to the *hashValue*, as shown in Figure 8-3.

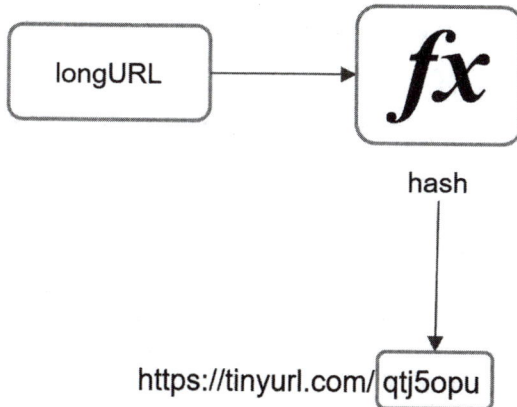

longURL → *fx*

hash

https://tinyurl.com/qtj5opu

Figure 8-3

The hash function must satisfy the following requirements:

- Each *longURL* must be hashed to one *hashValue*.
- Each *hashValue* can be mapped back to the *longURL*.

Detailed design for the hash function is discussed in deep dive.

Step 3 - Design deep dive

Up until now, we have discussed the high-level design of URL shortening and URL redirecting. In this section, we dive deep into the following: data model, hash function, URL shortening and URL redirecting.

Data model

In the high-level design, everything is stored in a hash table. This is a good starting point; however, this approach is not feasible for real-world systems as memory resources are limited and expensive. A better option is to store *<shortURL, longURL>* mapping in a relational database. Figure 8-4 shows a simple database table design. The simplified version of the table contains 3 columns: *id, shortURL, longURL*.

url	
PK	**id**
	shortURL
	longURL

Figure 8-4

Hash function

Hash function is used to hash a long URL to a short URL, also known as *hashValue*.

Hash value length

The *hashValue* consists of characters from [0-9, a-z, A-Z], containing 10 + 26 + 26 = 62 possible characters. To figure out the length of *hashValue*, find the smallest n such that $62^n \geq 365$ *billion*. The system must support up to 365 billion URLs based on the back of the envelope estimation. Table 8-1 shows the length of *hashValue* and the corresponding maximal number of URLs it can support.

n	Maximal number of URLs
1	$62^1 = 62$
2	$62^2 = 3,844$
3	$62^3 = 238,328$
4	$62^4 = 14,776,336$
5	$62^5 = 916,132,832$
6	$62^6 = 56,800,235,584$
7	$62^7 = 3,521,614,606,208 = $ ~3.5 trillion
8	$62^8 = 218,340,105,584,896$

Table 8-1

When $n = 7$, $62^n = $ ~3.5 *trillion*, 3.5 trillion is more than enough to hold 365 billion URLs, so the length of *hashValue* is 7.

We will explore two types of hash functions for a URL shortener. The first one is "hash + collision resolution", and the second one is "base 62 conversion." Let us look at them one by one.

Hash + collision resolution

To shorten a long URL, we should implement a hash function that hashes a long URL to a 7-character string. A straightforward solution is to use well-known hash functions like CRC32, MD5, or SHA-1. The following table compares the hash results after applying different hash functions on this URL: https://en.wikipedia.org/wiki/Systems_design.

Hash function	Hash value (Hexadecimal)
CRC32	5cb54054
MD5	5a62509a84df9ee03fe1230b9df8b84e
SHA-1	0eeae7916c06853901d9ccbefbfcaf4de57ed85b

Table 8-2

As shown in Table 8-2, even the shortest hash value (from CRC32) is too long (more than 7 characters). How can we make it shorter?

The first approach is to collect the first 7 characters of a hash value; however, this method can lead to hash collisions. To resolve hash collisions, we can recursively append a new predefined string until no more collision is discovered. This process is explained in Figure 8-5.

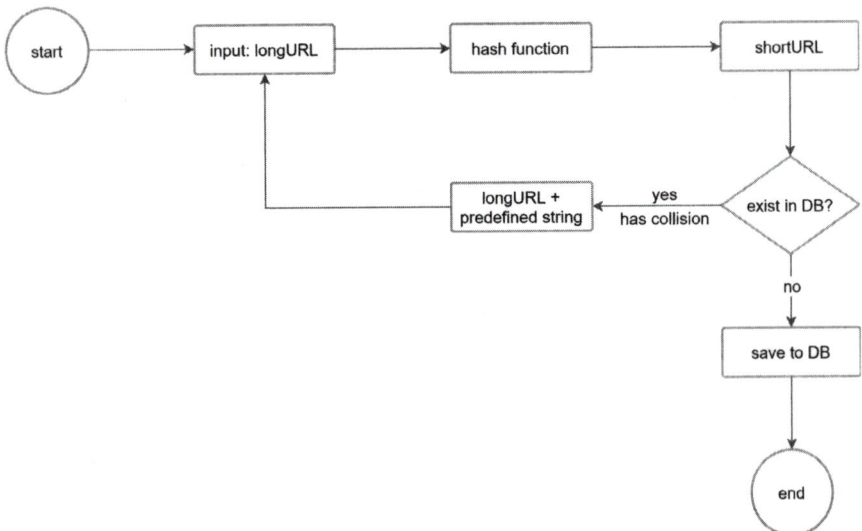

Figure 8-5

This method can eliminate collision; however, it is expensive to query the database to check if a shortURL exists for every request. A technique called bloom filters [2] can improve performance. A bloom filter is a space-efficient probabilistic technique to test if an element is a member of a set. Refer to the reference material [2] for more details.

Base 62 conversion

Base conversion is another approach commonly used for URL shorteners. Base conversion helps to convert the same number between its different number representation systems. Base 62 conversion is used as there are 62 possible characters for *hashValue*. Let us use an example to explain how the conversion works: convert 11157_{10} to base 62 representation (11157_{10} represents 11157 in a base 10 system).

- From its name, base 62 is a way of using 62 characters for encoding. The mappings are: *0-0, ..., 9-9, 10-a, 11-b, ..., 35-z, 36-A, ..., 61-Z, where 'a' stands for 10, 'Z' stands for 61, etc.*

- $11157_{10} = 2 \times 62^2 + 55 \times 62^1 + 59 \times 62^0 = [2, 55, 59] \rightarrow [2, T, X]$ in base 62 representation. Figure 8-6 shows the conversation process.

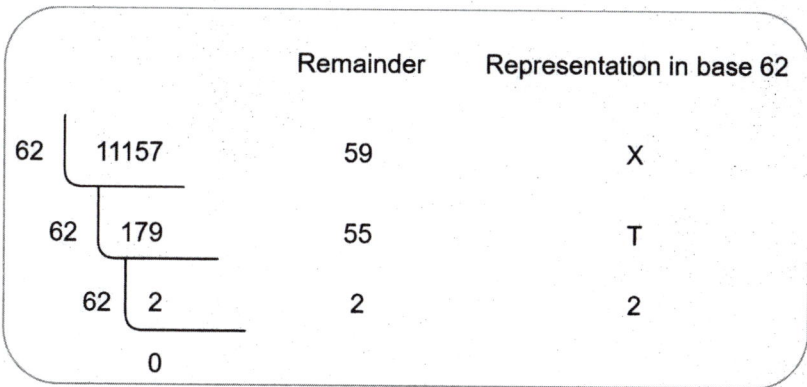

Figure 8-6

- Thus, the short URL is https://tinyurl.com **/2TX**

Comparison of the two approaches

Table 8-3 shows the differences of the two approaches.

Hash + collision resolution	Base 62 conversion
Fixed short URL length.	Short URL length is not fixed. It goes up with the ID.
Does not need a unique ID generator.	This option depends on a unique ID generator.
Collision is possible and needs to be resolved.	Collision is not possible because ID is unique.
It's not possible to figure out the next available short URL because it doesn't depend on ID.	It is easy to figure out what is the next available short URL if ID increments by 1 for a new entry. This can be a security concern.

Table 8-3

URL shortening deep dive

As one of the core pieces of the system, we want the URL shortening flow to be logically simple and functional. Base 62 conversion is used in our design. We build the following diagram (Figure 8-7) to demonstrate the flow.

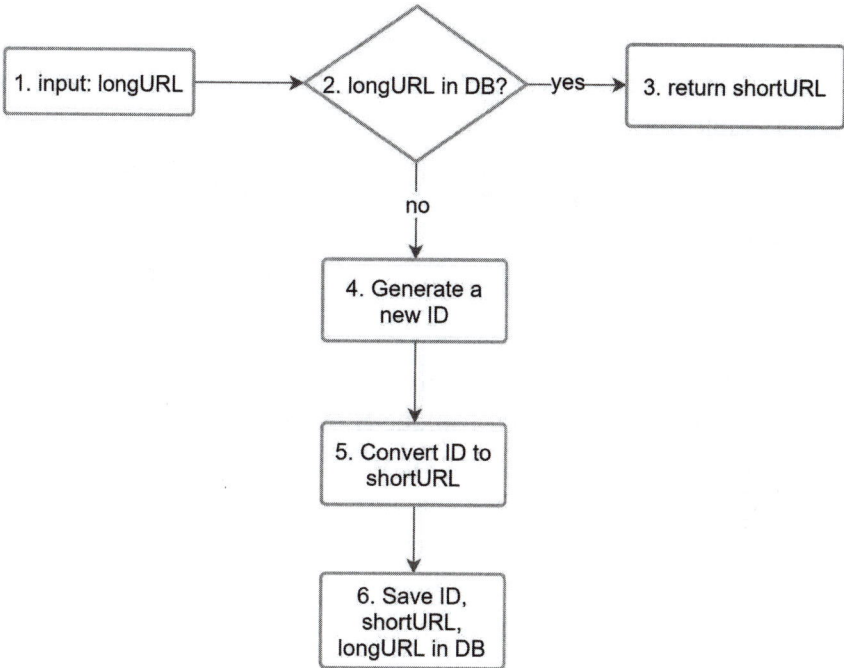

Figure 8-7

1. longURL is the input.

2. The system checks if the longURL is in the database.

3. If it is, it means the longURL was converted to shortURL before. In this case, fetch the shortURL from the database and return it to the client.

4. If not, the longURL is new. A new unique ID (primary key) Is generated by the unique ID generator.

5. Convert the ID to shortURL with base 62 conversion.

6. Create a new database row with the ID, shortURL, and longURL.

To make the flow easier to understand, let us look at a concrete example.

- Assuming the input longURL is: https://en.wikipedia.org/wiki/ Systems_design

- Unique ID generator returns ID: 2009215674938.

- Convert the ID to shortURL using the base 62 conversion. ID (2009215674938) is converted to "zn9edcu".
- Save ID, shortURL, and longURL to the database as shown in Table 8-4.

id	shortURL	longURL
2009215674938	zn9edcu	https://en.wikipedia.org/wiki/Systems_design

Table 8-4

The distributed unique ID generator is worth mentioning. Its primary function is to generate globally unique IDs, which are used for creating shortURLs. In a highly distributed environment, implementing a unique ID generator is challenging. Luckily, we have already discussed a few solutions in "Chapter 7: Design A Unique ID Generator in Distributed Systems". You can refer back to it to refresh your memory.

URL redirecting deep dive

Figure 8-8 shows the detailed design of the URL redirecting. As there are more reads than writes, *<shortURL, longURL>* mapping is stored in a cache to improve performance.

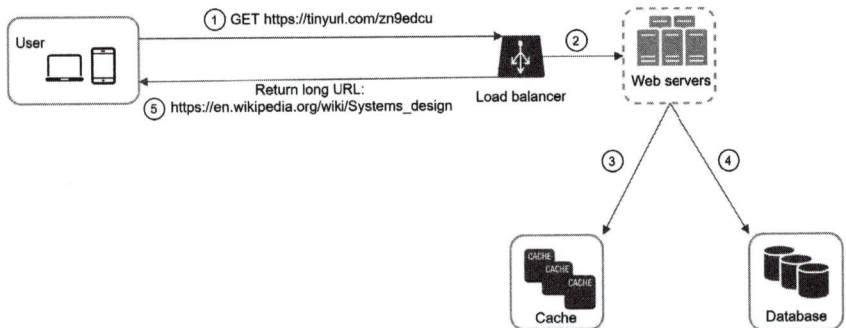

Figure 8-8

The flow of URL redirecting is summarized as follows:

1. A user clicks a short URL link: https://tinyurl.com/zn9edcu

2. The load balancer forwards the request to web servers.

3. If a shortURL is already in the cache, return the longURL directly.

4. If a shortURL is not in the cache, fetch the longURL from the database. If it is not in the database, it is likely a user entered an invalid shortURL.

5. The longURL is returned to the user.

Step 4 - Wrap up

In this chapter, we talked about the API design, data model, hash function, URL shortening, and URL redirecting.

If there is extra time at the end of the interview, here are a few additional talking points.

- Rate limiter: A potential security problem we could face is that malicious users send an overwhelmingly large number of URL shortening requests. Rate limiter helps to filter out requests based on IP address or other filtering rules. If you want to refresh your memory about rate limiting, refer to "Chapter 4: Design a rate limiter".

- Web server scaling: Since the web tier is stateless, it is easy to scale the web tier by adding or removing web servers.

- Database scaling: Database replication and sharding are common techniques.

- Analytics: Data is increasingly important for business success. Integrating an analytics solution to the URL shortener could help to answer important questions like how many people click on a link? When do they click the link? etc.

- Availability, consistency, and reliability. These concepts are at the core of any large system's success. We discussed them in detail in Chapter 1, please refresh your memory on these topics.

Congratulations on getting this far! Now give yourself a pat on the back. Good job!

Reference materials

[1] A RESTful Tutorial: https://www.restapitutorial.com/index.html

[2] Bloom filter: https://en.wikipedia.org/wiki/Bloom_filter

9

DESIGN A WEB CRAWLER

In this chapter, we focus on web crawler design: an interesting and classic system design interview question.

A web crawler is known as a robot or spider. It is widely used by search engines to discover new or updated content on the web. Content can be a web page, an image, a video, a PDF file, etc. A web crawler starts by collecting a few web pages and then follows links on those pages to collect new content. Figure 9-1 shows a visual example of the crawl process.

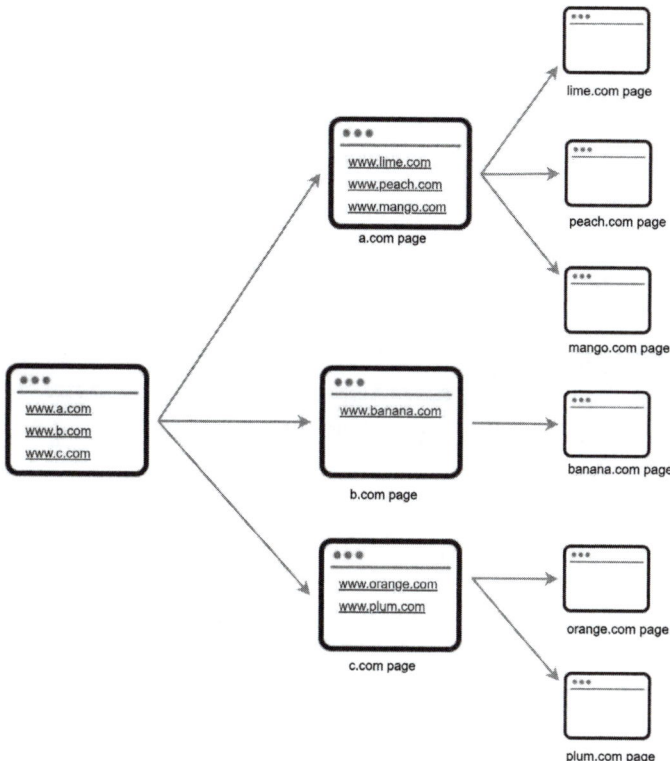

Figure 9-1

A crawler is used for many purposes:

- Search engine indexing: This is the most common use case. A crawler collects web pages to create a local index for search engines. For example, Googlebot is the web crawler behind the Google search engine.

- Web archiving: This is the process of collecting information from the web to preserve data for future uses. For instance, many national libraries run crawlers to archive web sites. Notable examples are the US Library of Congress [1] and the EU web archive [2].

- Web mining: The explosive growth of the web presents an unprecedented opportunity for data mining. Web mining helps to discover useful knowledge from the internet. For example, top financial firms use crawlers to download shareholder meetings and annual reports to learn key company initiatives.

- Web monitoring. The crawlers help to monitor copyright and trademark infringements over the Internet. For example, Digimarc [3] utilizes crawlers to discover pirated works and reports.

The complexity of developing a web crawler depends on the scale we intend to support. It could be either a small school project, which takes only a few hours to complete or a gigantic project that requires continuous improvement from a dedicated engineering team. Thus, we will explore the scale and features to support below.

Step 1 - Understand the problem and establish design scope

The basic algorithm of a web crawler is simple:

1. Given a set of URLs, download all the web pages addressed by the URLs.

2. Extract URLs from these web pages

3. Add new URLs to the list of URLs to be downloaded. Repeat these 3 steps.

Does a web crawler work truly as simple as this basic algorithm? Not exactly. Designing a vastly scalable web crawler is an extremely complex task. It is unlikely for anyone to design a massive web crawler within the interview duration. Before jumping into the design, we must ask questions to understand the requirements and establish design scope:

> **Candidate**: What is the main purpose of the crawler? Is it used for search engine indexing, data mining, or something else?
> **Interviewer**: Search engine indexing.

> **Candidate**: How many web pages does the web crawler collect per month?
> **Interviewer**: 1 billion pages.

> **Candidate**: What content types are included? HTML only or other content types such as PDFs and images as well?
> **Interviewer**: HTML only.

> **Candidate**: Shall we consider newly added or edited web pages?
> **Interviewer**: Yes, we should consider the newly added or edited web pages.

> **Candidate**: Do we need to store HTML pages crawled from the web?
> **Interviewer**: Yes, up to 5 years

> **Candidate**: How do we handle web pages with duplicate content?
> **Interviewer**: Pages with duplicate content should be ignored.

Above are some of the sample questions that you can ask your interviewer. It is important to understand the requirements and clarify ambiguities. Even if you are asked to design a straightforward product like a web crawler, you and your interviewer might not have the same assumptions.

Beside functionalities to clarify with your interviewer, it is also important to note down the following characteristics of a good web crawler:

- Scalability: The web is very large. There are billions of web pages out there. Web crawling should be extremely efficient using parallelization.

- Robustness: The web is full of traps. Bad HTML, unresponsive servers, crashes, malicious links, etc. are all common. The crawler must handle all those edge cases.

- Politeness: The crawler should not make too many requests to a website within a short time interval.

- Extensibility: The system is flexible so that minimal changes are needed to support new content types. For example, if we want to crawl image files in the future, we should not need to redesign the entire system.

Back of the envelope estimation

The following estimations are based on many assumptions, and it is important to communicate with the interviewer to be on the same page.

- Assume 1 billion web pages are downloaded every month.

- QPS: 1,000,000,000 / 30 days / 24 hours / 3600 seconds = ~400 pages per second.

- Peak QPS = 2 * QPS = 800

- Assume the average web page size is 500k.

- 1-billion-page x 500k = 500 TB storage per month. If you are unclear about digital storage units, go through "Power of 2" section in Chapter 2 again.

- Assuming data are stored for five years, 500 TB * 12 months * 5 years = 30 PB. A 30 PB storage is needed to store five-year content.

Step 2 - Propose high-level design and get buy-in

Once the requirements are clear, we move on to the high-level design. Inspired by previous studies on web crawling [4] [5], we propose a high-level design as shown in Figure 9-2.

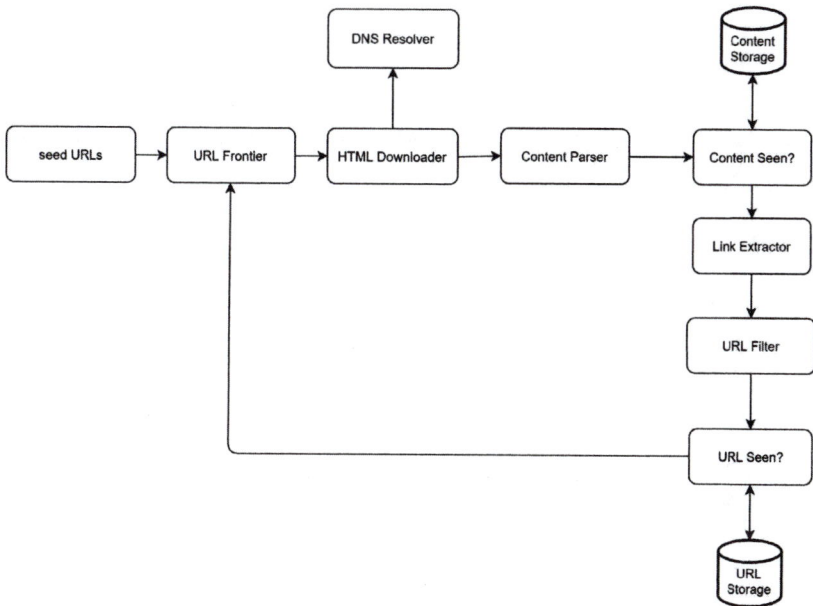

Figure 9-2

First, we explore each design component to understand their functionalities. Then, we examine the crawler workflow step-by-step.

Seed URLs

A web crawler uses seed URLs as a starting point for the crawl process. For example, to crawl all web pages from a university's website, an intuitive way to select seed URLs is to use the university's domain name.

To crawl the entire web, we need to be creative in selecting seed URLs. A good seed URL serves as a good starting point that a crawler can utilize to traverse as many links as possible. The general strategy is to divide the entire URL space into smaller ones. The first proposed approach is based on locality as different countries may have different popular websites. Another way is to choose seed URLs based on topics; for example, we can divide URL space into shopping, sports, healthcare, etc. Seed URL selection is an open-ended question. You are not expected to give the perfect answer. Just think out loud.

URL Frontier

Most modern web crawlers split the crawl state into two: to be downloaded and already downloaded. The component that stores URLs to be downloaded is called the URL Frontier. You can refer to this as a First-in-First-out (FIFO) queue. For detailed information about the URL Frontier, refer to the deep dive.

HTML Downloader

The HTML downloader downloads web pages from the internet. Those URLs are provided by the URL Frontier.

DNS Resolver

To download a web page, a URL must be translated into an IP address. The HTML Downloader calls the DNS Resolver to get the corresponding IP address for the URL. For instance, URL www.wikipedia.org is converted to IP address 198.35.26.96 as of 3/5/2019.

Content Parser

After a web page is downloaded, it must be parsed and validated because malformed web pages could provoke problems and waste storage space. Implementing a content parser in a crawl server will slow down the crawling process. Thus, the content parser is a separate component.

Content Seen?

Online research [6] reveals that 29% of the web pages are duplicated contents, which may cause the same content to be stored multiple times. We introduce the "Content Seen?" data structure to eliminate data redundancy and shorten processing time. It helps to detect new content previously stored in the system. To compare two HTML documents, we can compare them character by character. However, this method is slow and time-consuming, especially when billions of web pages are involved. An efficient way to accomplish this task is to compare the hash values of the two web pages [7].

Content Storage

It is a storage system for storing HTML content. The choice of storage system depends on factors such as data type, data size, access frequency, life span, etc. Both disk and memory are used.

- Most of the content is stored on disk because the data set is too big to fit in memory.

- Popular content is kept in memory to reduce latency.

URL Extractor

URL Extractor parses and extracts links from HTML pages. Figure 9-3 shows an example of a link extraction process. Relative paths are converted to absolute URLs by adding the "https://en.wikipedia.org" prefix.

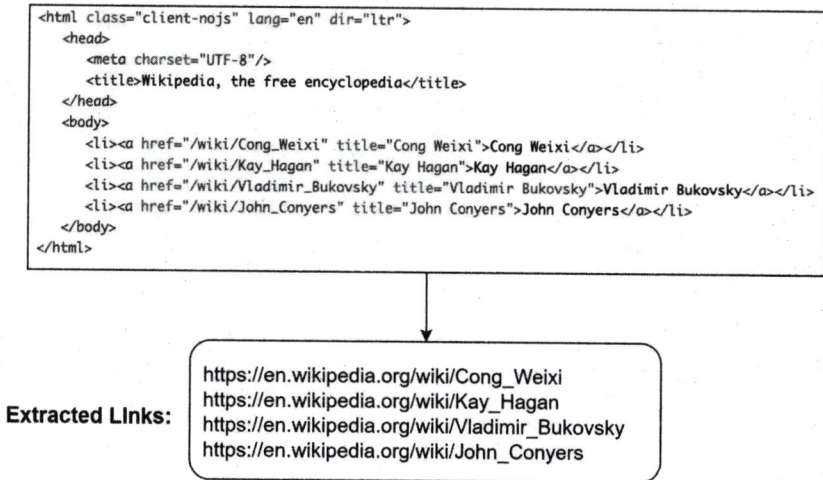

```
<html class="client-nojs" lang="en" dir="ltr">
    <head>
        <meta charset="UTF-8"/>
        <title>Wikipedia, the free encyclopedia</title>
    </head>
    <body>
        <li><a href="/wiki/Cong_Weixi" title="Cong Weixi">Cong Weixi</a></li>
        <li><a href="/wiki/Kay_Hagan" title="Kay Hagan">Kay Hagan</a></li>
        <li><a href="/wiki/Vladimir_Bukovsky" title="Vladimir Bukovsky">Vladimir Bukovsky</a></li>
        <li><a href="/wiki/John_Conyers" title="John Conyers">John Conyers</a></li>
    </body>
</html>
```

Extracted Links:
```
https://en.wikipedia.org/wiki/Cong_Weixi
https://en.wikipedia.org/wiki/Kay_Hagan
https://en.wikipedia.org/wiki/Vladimir_Bukovsky
https://en.wikipedia.org/wiki/John_Conyers
```

Figure 9-3

URL Filter

The URL filter excludes certain content types, file extensions, error links and URLs in "blacklisted" sites.

URL Seen?

"URL Seen?" is a data structure that keeps track of URLs that are visited before or already in the Frontier. "URL Seen?" helps to avoid adding the same URL multiple times as this can increase server load and cause potential infinite loops.

Bloom filter and hash table are common techniques to implement the "URL Seen?" component. We will not cover the detailed implementation of the bloom filter and hash table here. For more information, refer to the reference materials [4] [8].

URL Storage

URL Storage stores already visited URLs.

So far, we have discussed every system component. Next, we put them together to explain the workflow.

Web crawler workflow

To better explain the workflow step-by-step, sequence numbers are added in the design diagram as shown in Figure 9-4.

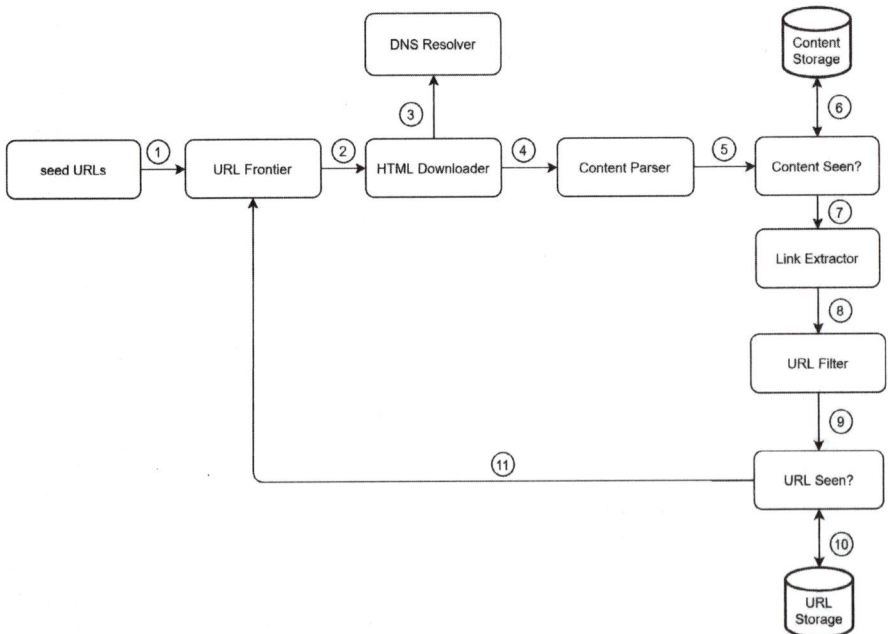

Figure 9-4

Step 1: Add seed URLs to the URL Frontier

Step 2: HTML Downloader fetches a list of URLs from URL Frontier.

Step 3: HTML Downloader gets IP addresses of URLs from DNS resolver and starts downloading.

Step 4: Content Parser parses HTML pages and checks if pages are malformed.

Step 5: After content is parsed and validated, it is passed to the "Content Seen?" component.

Step 6: "Content Seen" component checks if a HTML page is already in the storage.

- If it is in the storage, this means the same content in a different URL has already been processed. In this case, the HTML page is discarded.

- If it is not in the storage, the system has not processed the same content before. The content is passed to Link Extractor.

Step 7: Link extractor extracts links from HTML pages.

Step 8: Extracted links are passed to the URL filter.

Step 9: After links are filtered, they are passed to the "URL Seen?" component.

Step 10: "URL Seen" component checks if a URL is already in the storage, if yes, it is processed before, and nothing needs to be done.

Step 11: If a URL has not been processed before, it is added to the URL Frontier.

Step 3 - Design deep dive

Up until now, we have discussed the high-level design. Next, we will discuss the most important building components and techniques in depth:

- Depth-first search (DFS) vs Breadth-first search (BFS)

- URL frontier
- HTML Downloader
- Robustness
- Extensibility
- Detect and avoid problematic content

DFS vs BFS

You can think of the web as a directed graph where web pages serve as nodes and hyperlinks (URLs) as edges. The crawl process can be seen as traversing a directed graph from one web page to others. Two common graph traversal algorithms are DFS and BFS. However, DFS is usually not a good choice because the depth of DFS can be very deep.

BFS is commonly used by web crawlers and is implemented by a first-in-first-out (FIFO) queue. In a FIFO queue, URLs are dequeued in the order they are enqueued. However, this implementation has two problems:

- Most links from the same web page are linked back to the same host. In Figure 9-5, all the links in wikipedia.com are internal links, making the crawler busy processing URLs from the same host (wikipedia.com). When the crawler tries to download web pages in parallel, Wikipedia servers will be flooded with requests. This is considered as "impolite".

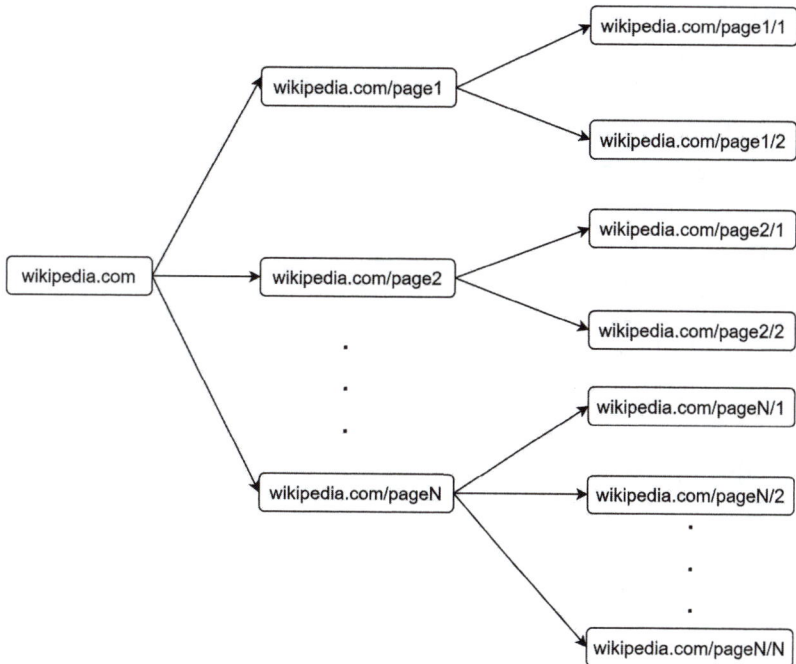

Figure 9-5

- Standard BFS does not take the priority of a URL into consideration. The web is large and not every page has the same level of quality and importance. Therefore, we may want to prioritize URLs according to their page ranks, web traffic, update frequency, etc.

URL frontier

URL frontier helps to address these problems. A URL frontier is a data structure that stores URLs to be downloaded. The URL frontier is an important component to ensure politeness, URL prioritization, and freshness. A few noteworthy papers on URL frontier are mentioned in the reference materials [5] [9]. The findings from these papers are as follows:

Politeness

Generally, a web crawler should avoid sending too many requests to the same hosting server within a short period. Sending too many requests is considered as "impolite" or even treated as denial-of-service (DOS) attack. For example, without any constraint, the crawler can send thousands of requests every second to the same website. This can overwhelm the web servers.

The general idea of enforcing politeness is to download one page at a time from the same host. A delay can be added between two download tasks. The politeness constraint is implemented by maintain a mapping from website hostnames to download (worker) threads. Each downloader thread has a separate FIFO queue and only downloads URLs obtained from that queue. Figure 9-6 shows the design that manages politeness.

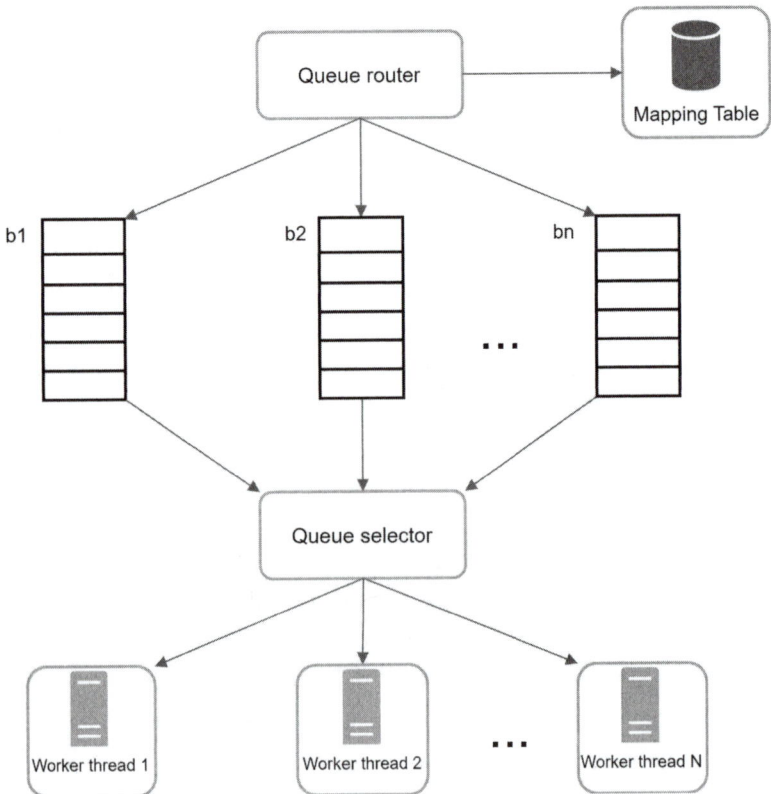

Figure 9-6

- Queue router: It ensures that each queue (b1, b2, ... bn) only contains URLs from the same host.

- Mapping table: It maps each host to a queue.

Host	Queue
wikipedia.com	b1
apple.com	b2
...	...
nike.com	bn

Table 9-1

- FIFO queues b1, b2 to bn: Each queue contains URLs from the same host.

- Queue selector: Each worker thread is mapped to a FIFO queue, and it only downloads URLs from that queue. The queue selection logic is done by the Queue selector.

- Worker thread 1 to N. A worker thread downloads web pages one by one from the same host. A delay can be added between two download tasks.

Priority

A random post from a discussion forum about Apple products carries very different weight than posts on the Apple home page. Even though they both have the "Apple" keyword, it is sensible for a crawler to crawl the Apple home page first.

We prioritize URLs based on usefulness, which can be measured by PageRank [10], website traffic, update frequency, etc. "Prioritizer" is the component that handles URL prioritization. Refer to the reference materials [5] [10] for in-depth information about this concept.

Figure 9-7 shows the design that manages URL priority.

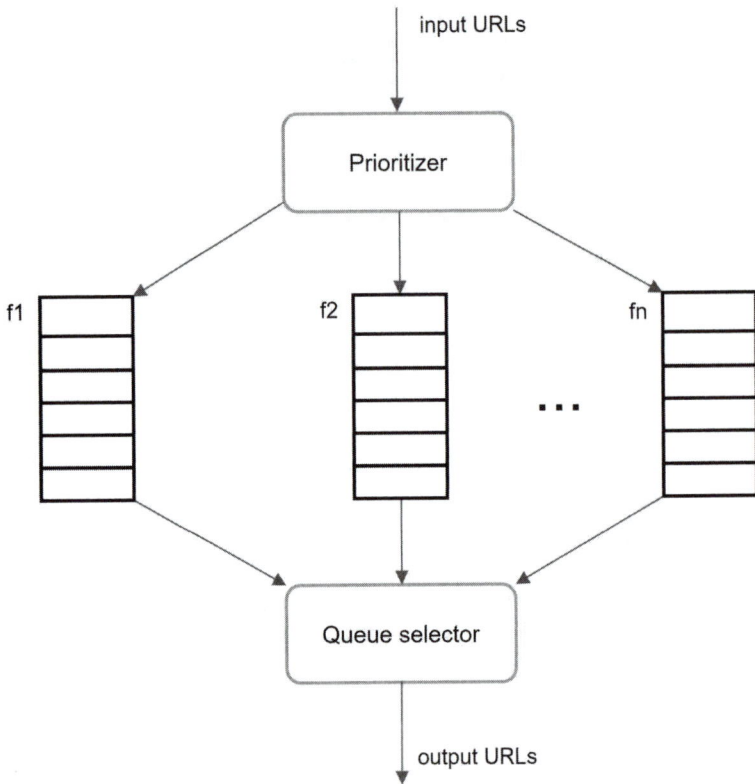

Figure 9-7

- Prioritizer: It takes URLs as input and computes the priorities.

- Queue f1 to fn: Each queue has an assigned priority. Queues with high priority are selected with higher probability.

- Queue selector: Randomly choose a queue with a bias towards queues with higher priority.

Figure 9-8 presents the URL frontier design, and it contains two modules:

- Front queues: manage prioritization

- Back queues: manage politeness

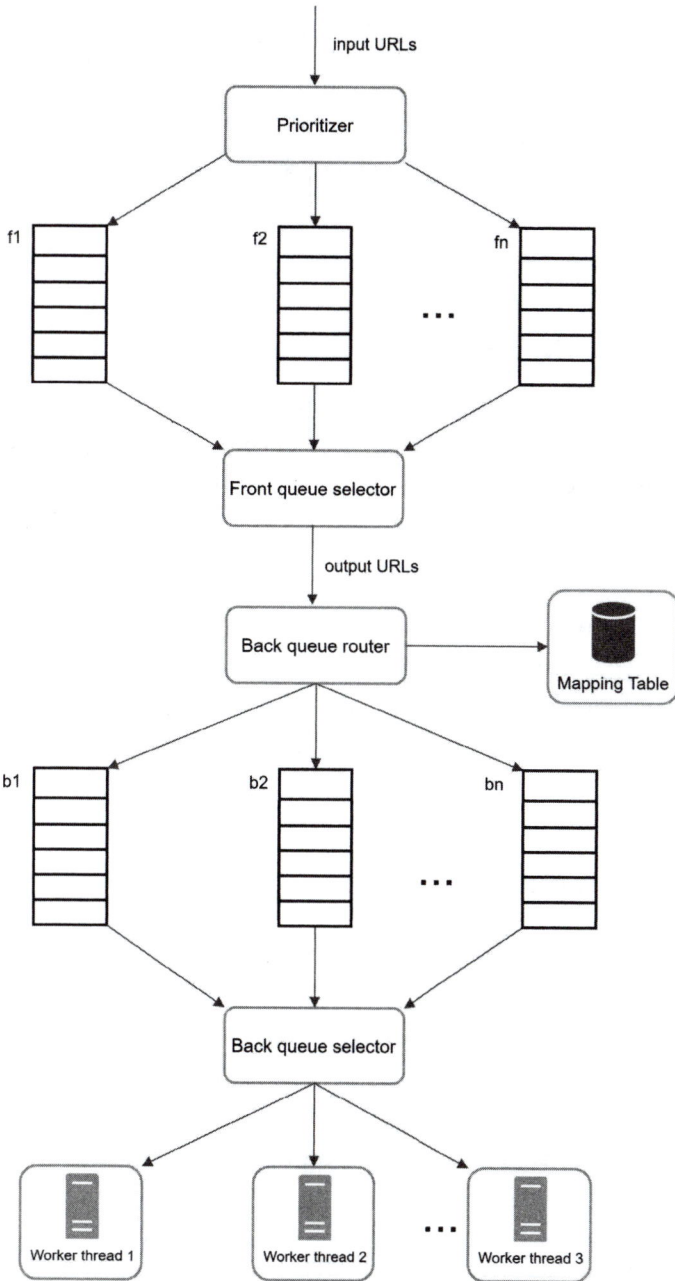

Figure 9-8

Freshness

Web pages are constantly being added, deleted, and edited. A web crawler must periodically recrawl downloaded pages to keep our data set fresh. Recrawl all the URLs is time-consuming and resource intensive. Few strategies to optimize freshness are listed as follows:

- Recrawl based on web pages' update history.
- Prioritize URLs and recrawl important pages first and more frequently.

Storage for URL Frontier

In real-world crawl for search engines, the number of URLs in the frontier could be hundreds of millions [4]. Putting everything in memory is neither durable nor scalable. Keeping everything in the disk is undesirable neither because the disk is slow; and it can easily become a bottleneck for the crawl.

We adopted a hybrid approach. The majority of URLs are stored on disk, so the storage space is not a problem. To reduce the cost of reading from the disk and writing to the disk, we maintain buffers in memory for enqueue/dequeue operations. Data in the buffer is periodically written to the disk.

HTML Downloader

The HTML Downloader downloads web pages from the internet using the HTTP protocol. Before discussing the HTML Downloader, we look at Robots Exclusion Protocol first.

Robots.txt

Robots.txt, called Robots Exclusion Protocol, is a standard used by websites to communicate with crawlers. It specifies what pages crawlers are allowed to download. Before attempting to crawl a web site, a crawler should check its corresponding robots.txt first and follow its rules.

To avoid repeat downloads of robots.txt file, we cache the results of the file. The file is downloaded and saved to cache periodically. Here is a piece of robots.txt file taken from https://www.amazon.com/robots.txt. Some of the directories like creatorhub are disallowed for Google bot.

User-agent: Googlebot
Disallow: /creatorhub/*
Disallow: /rss/people/*/reviews
Disallow: /gp/pdp/rss/*/reviews
Disallow: /gp/cdp/member-reviews/
Disallow: /gp/aw/cr/

Besides robots.txt, performance optimization is another important concept we will cover for the HTML downloader.

Performance optimization

Below is a list of performance optimizations for HTML downloader.

1. Distributed crawl

To achieve high performance, crawl jobs are distributed into multiple servers, and each server runs multiple threads. The URL space is partitioned into smaller pieces; so, each downloader is responsible for a subset of the URLs. Figure 9-9 shows an example of a distributed crawl.

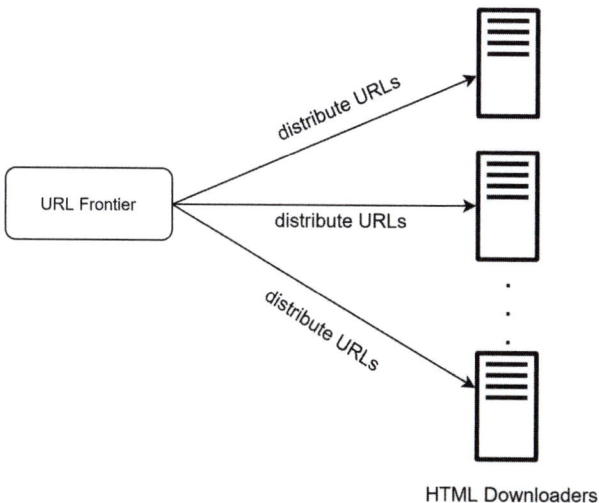

Figure 9-9

2. Cache DNS Resolver

DNS Resolver is a bottleneck for crawlers because DNS requests might take time due to the synchronous nature of many DNS interfaces. DNS response time ranges from 10ms to 200ms. Once a request to DNS is carried out by a crawler thread, other threads are blocked until the first request is completed. Maintaining our DNS cache to avoid calling DNS frequently is an effective technique for speed optimization. Our DNS cache keeps the domain name to IP address mapping and is updated periodically by cron jobs.

3. Locality

Distribute crawl servers geographically. When crawl servers are closer to website hosts, crawlers experience faster download time. Design locality applies to most of the system components: crawl servers, cache, queue, storage, etc.

4. Short timeout

Some web servers respond slowly or may not respond at all. To avoid long wait time, a maximal wait time is specified. If a host does not respond within a predefined time, the crawler will stop the job and crawl some other pages.

Robustness

Besides performance optimization, robustness is also an important consideration. We present a few approaches to improve the system robustness:

- Consistent hashing: This helps to distribute loads among downloaders. A new downloader server can be added or removed using consistent hashing. Refer to Chapter 5: Design consistent hashing for more details.

- Save crawl states and data: To guard against failures, crawl states and data are written to a storage system. A disrupted crawl can be restarted easily by loading saved states and data.

- Exception handling: Errors are inevitable and common in a

large-scale system. The crawler must handle exceptions gracefully without crashing the system.

- Data validation: This is an important measure to prevent system errors.

Extensibility

As almost every system evolves, one of the design goals is to make the system flexible enough to support new content types. The crawler can be extended by plugging in new modules. Figure 9-10 shows how to add new modules.

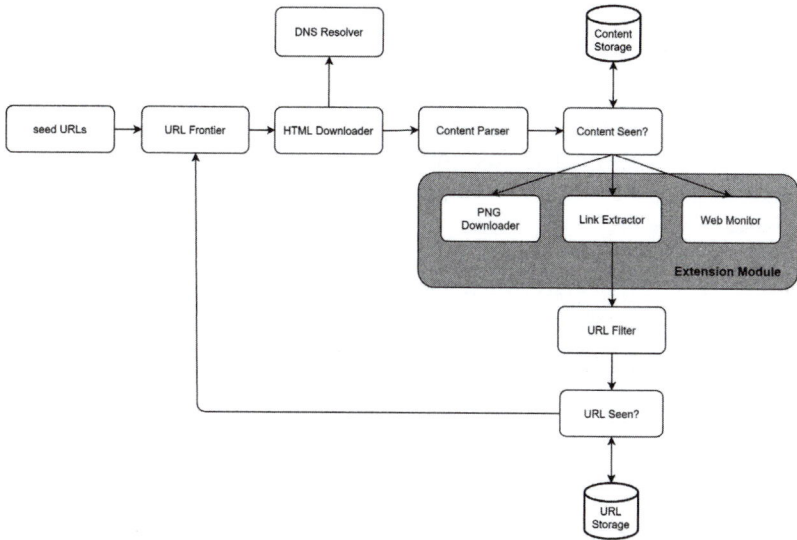

Figure 9-10

- PNG Downloader module is plugged-in to download PNG files.
- Web Monitor module is added to monitor the web and prevent copyright and trademark infringements.

Detect and avoid problematic content

This section discusses the detection and prevention of redundant, meaningless, or harmful content.

1. Redundant content

As discussed previously, nearly 30% of the web pages are duplicates. Hashes or checksums help to detect duplication [11].

2. Spider traps

A spider trap is a web page that causes a crawler in an infinite loop. For instance, an infinite deep directory structure is listed as follows: _www. spidertrapexample.com/foo/bar/foo/bar/foo/bar/..._

Such spider traps can be avoided by setting a maximal length for URLs. However, no one-size-fits-all solution exists to detect spider traps. Websites containing spider traps are easy to identify due to an unusually large number of web pages discovered on such websites. It is hard to develop automatic algorithms to avoid spider traps; however, a user can manually verify and identify a spider trap, and either exclude those websites from the crawler or apply some customized URL filters.

3. Data noise

Some of the contents have little or no value, such as advertisements, code snippets, spam URLs, etc. Those contents are not useful for crawlers and should be excluded if possible.

Step 4 - Wrap up

In this chapter, we first discussed the characteristics of a good crawler: scalability, politeness, extensibility, and robustness. Then, we proposed a design and discussed key components. Building a scalable web crawler is not a trivial task because the web is enormously large and full of traps. Even though we have covered many topics, we still miss many relevant talking points:

- Server-side rendering: Numerous websites use scripts like JavaScript, AJAX, etc to generate links on the fly. If we download and parse web pages directly, we will not be able to retrieve dynamically generated links. To solve this problem, we perform server-side rendering (also called dynamic rendering) first before parsing a page [12].

- Filter out unwanted pages: With finite storage capacity and crawl resources, an anti-spam component is beneficial in filtering out low quality and spam pages [13] [14].

- Database replication and sharding: Techniques like replication and sharding are used to improve the data layer availability, scalability, and reliability.

- Horizontal scaling: For large scale crawl, hundreds or even thousands of servers are needed to perform download tasks. The key is to keep servers stateless.

- Availability, consistency, and reliability: These concepts are at the core of any large system's success. We discussed these concepts in detail in Chapter 1. Refresh your memory on these topics.

- Analytics: Collecting and analyzing data are important parts of any system because data is key ingredient for fine-tuning.

Congratulations on getting this far! Now give yourself a pat on the back. Good job!

168   SYSTEM DESIGN INTERVIEW

Reference materials

[1] US Library of Congress: https://www.loc.gov/websites/

[2] EU Web Archive: http://data.europa.eu/webarchive

[3] Digimarc:
https://www.digimarc.com/products/digimarc-services/piracy-
intelligence

[4] Heydon A., Najork M. Mercator: A scalable, extensible web crawler
World Wide Web, 2 (4) (1999), pp. 219-229

[5] By Christopher Olston, Marc Najork: Web Crawling.
http://infolab.stanford.edu/~olston/publications/crawling_survey.pdf

[6] 29% Of Sites Face Duplicate Content Issues:
https://tinyurl.com/y6tmh55y

[7] Rabin M.O., et al. Fingerprinting by random polynomials Center
for Research in Computing Techn., Aiken Computation Laboratory,
Univ. (1981)

[8] B. H. Bloom, "Space/time trade-offs in hash coding with
allowable errors," Communications of the ACM, vol. 13, no. 7, pp.
422–426, 1970.

[9] Donald J. Patterson, Web Crawling:
https://www.ics.uci.edu/~lopes/teaching/cs221W12/slides/
Lecture05.pdf

[10] L. Page, S. Brin, R. Motwani, and T. Winograd, "The PageRank
citation ranking: Bringing order to the web," Technical Report,
Stanford University, 1998.

[11] Burton Bloom. Space/time trade-offs in hash coding with
allowable errors. Communications of the ACM, 13(7), pages 422--426,
July 1970.

[12] Google Dynamic Rendering:
https://developers.google.com/search/docs/guides/dynamic-rendering

[13] T. Urvoy, T. Lavergne, and P. Filoche, "Tracking web spam with hidden style similarity," in Proceedings of the 2nd International Workshop on Adversarial Information Retrieval on the Web, 2006.

[14] H.-T. Lee, D. Leonard, X. Wang, and D. Loguinov, "IRLbot: Scaling to 6 billion pages and beyond," in Proceedings of the 17th International World Wide Web Conference, 2008.

10

DESIGN A NOTIFICATION SYSTEM

A notification system has already become a very popular feature for many applications in recent years. A notification alerts a user with important information like breaking news, product updates, events, offerings, etc. It has become an indispensable part of our daily life. In this chapter, you are asked to design a notification system.

A notification is more than just mobile push notification. Three types of notification formats are: mobile push notification, SMS message, and Email. Figure 10-1 shows an example of each of these notifications.

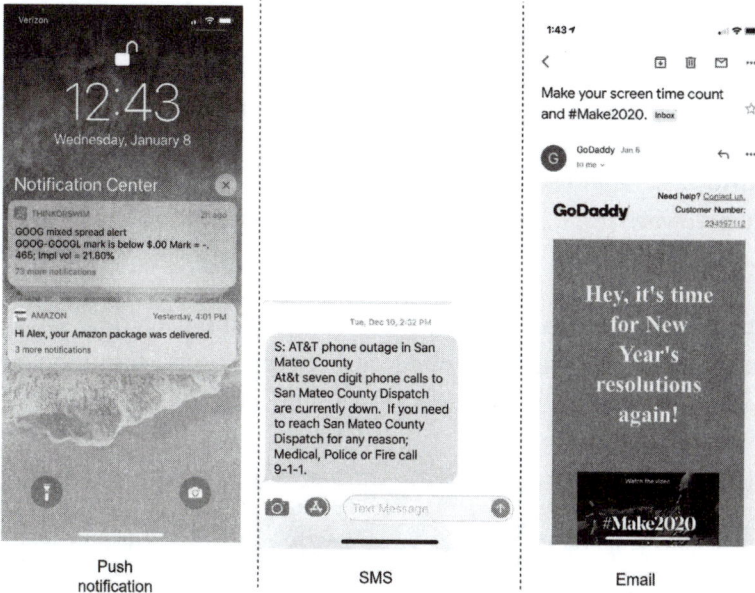

Push notification

SMS

Email

Figure 10-1

Step 1 - Understand the problem and establish design scope

Building a scalable system that sends out millions of notifications a day is not an easy task. It requires a deep understanding of the notification ecosystem. The interview question is purposely designed to be open-ended and ambiguous, and it is your responsibility to ask questions to clarify the requirements.

Candidate: What types of notifications does the system support?
Interviewer: Push notification, SMS message, and email.

Candidate: Is it a real-time system?
Interviewer: Let us say it is a soft real-time system. We want a user to receive notifications as soon as possible. However, if the system is under a high workload, a slight delay is acceptable.

Candidate: What are the supported devices?
Interviewer: iOS devices, android devices, and laptop/desktop.

Candidate: What triggers notifications?
Interviewer: Notifications can be triggered by client applications. They can also be scheduled on the server-side.

Candidate: Will users be able to opt-out?
Interviewer: Yes, users who choose to opt-out will no longer receive notifications.

Candidate: How many notifications are sent out each day?
Interviewer: 10 million mobile push notifications, 1 million SMS messages, and 5 million emails.

Step 2 - Propose high-level design and get buy-in

This section shows the high-level design that supports various notification types: iOS push notification, Android push notification, SMS message, and Email. It is structured as follows:

- Different types of notifications
- Contact info gathering flow
- Notification sending/receiving flow

Different types of notifications

We start by looking at how each notification type works at a high level.

iOS push notification

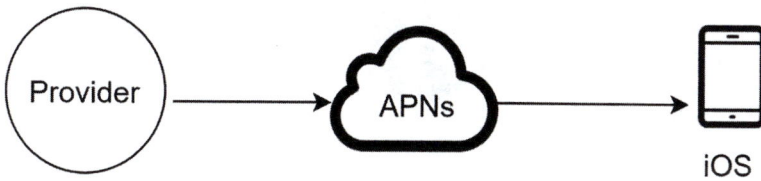

Figure 10-2

We primary need three components to send an iOS push notification:

- Provider. A provider builds and sends notification requests to Apple Push Notification Service (APNS). To construct a push notification, the provider provides the following data:

  o Device token: This is a unique identifier used for sending push notifications.

  o Payload: This is a JSON dictionary that contains a notification's payload. Here is an example:

```
{
    "aps" : {
            "alert" : {
                "title" : "Game Request",
                "body" : "Bob wants to play chess",
                "action-loc-key" : "PLAY"
            },
            "badge" : 5
    }
}
```

- APNS: This is a remote service provided by Apple to propagate push notifications to iOS devices.

- iOS Device: It is the end client, which receives push notifications.

Android push notification

Android adopts a similar notification flow. Instead of using APNs, Firebase Cloud Messaging (FCM) is commonly used to send push notifications to android devices.

Figure 10-3

SMS message

For SMS messages, third party SMS services like Twilio [1], Nexmo [2], and many others are commonly used. Most of them are commercial services.

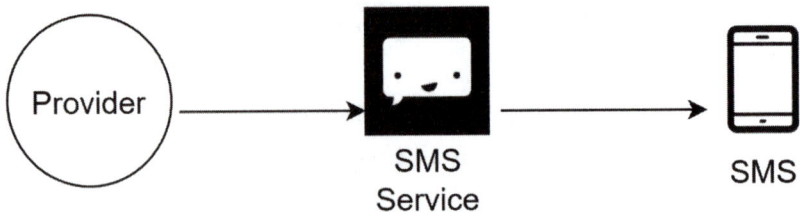

Figure 10-4

Email

Although companies can set up their own email servers, many of them opt for commercial email services. Sendgrid [3] and Mailchimp [4] are among the most popular email services, which offer a better delivery rate and data analytics.

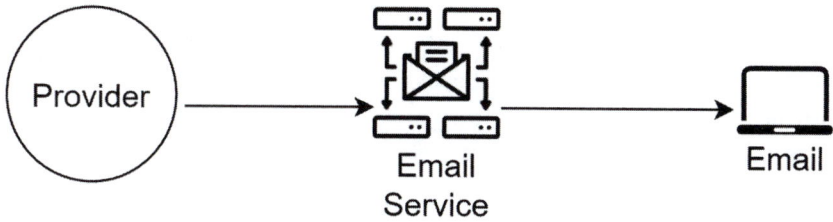

Figure 10-5

Figure 10-6 shows the design after including all the third-party services.

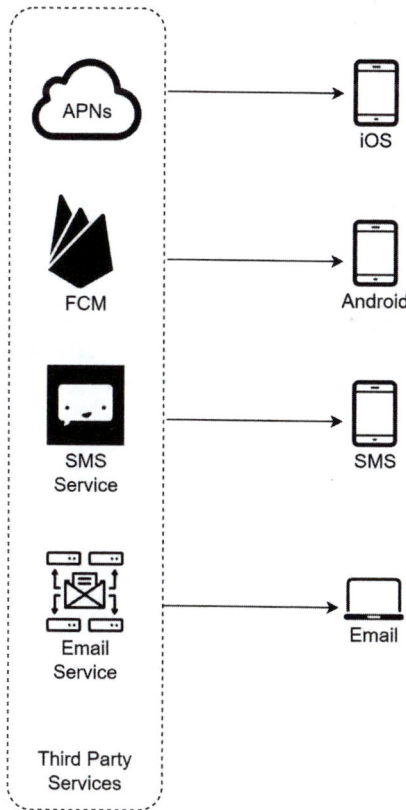

Figure 10-6

Contact info gathering flow

To send notifications, we need to gather mobile device tokens, phone numbers, or email addresses. As shown in Figure 10-7, when a user installs our app or signs up for the first time, API servers collect user contact info and store it in the database.

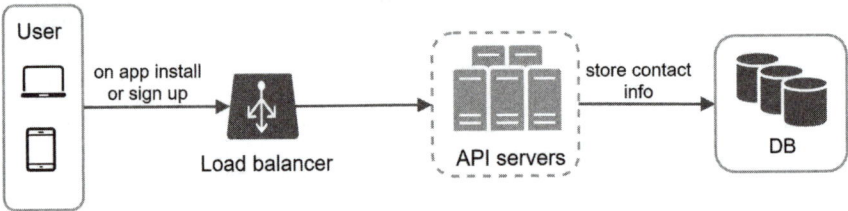

Figure 10-7

Figure 10-8 shows simplified database tables to store contact info. Email addresses and phone numbers are stored in the *user* table, whereas device tokens are stored in the *device* table. A user can have multiple devices, indicating that a push notification can be sent to all the user devices.

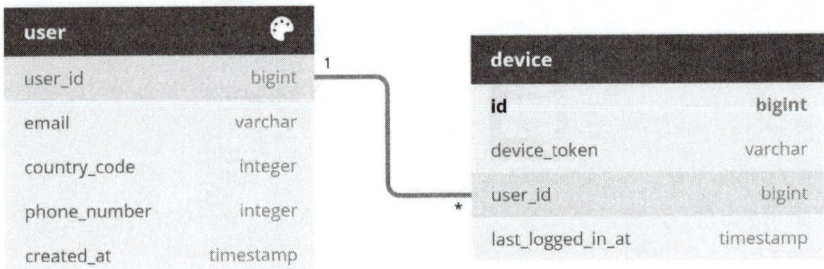

Figure 10-8

Notification sending/receiving flow

We will first present the initial design; then, propose some optimizations.

High-level design

Figure 10-9 shows the design, and each system component is explained below.

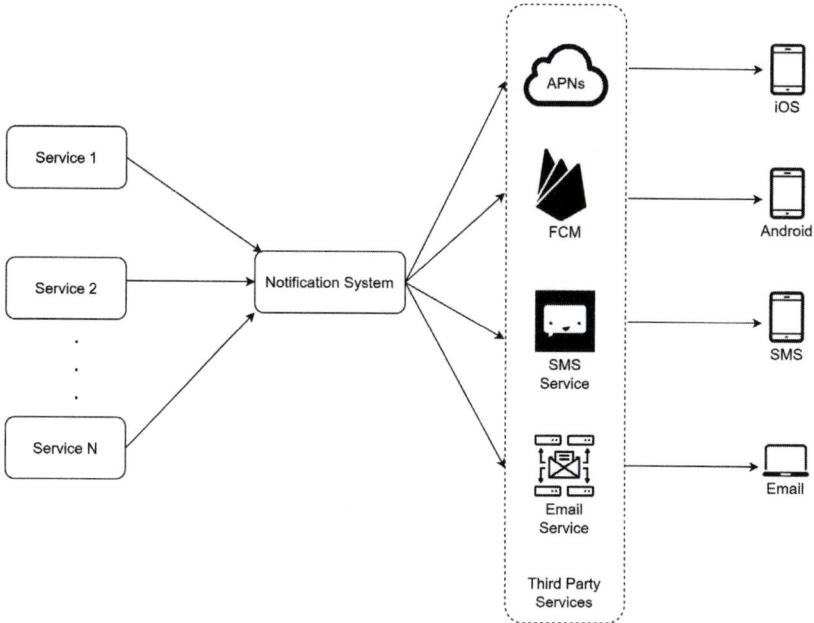

Figure 10-9

Service 1 to N: A service can be a micro-service, a cron job, or a distributed system that triggers notification sending events. For example, a billing service sends emails to remind customers of their due payment or a shopping website tells customers that their packages will be delivered tomorrow via SMS messages.

Notification system: The notification system is the centerpiece of sending/receiving notifications. Starting with something simple, only one notification server is used. It provides APIs for services 1 to N, and builds notification payloads for third party services.

Third-party services: Third party services are responsible for delivering notifications to users. While integrating with third-party services, we need to pay extra attention to extensibility. Good extensibility means a flexible system that can easily plugging or unplugging of a third-party service. Another important consideration is that a third-party service might be unavailable in new markets or in the future. For instance, FCM is unavailable in China. Thus, alternative third-party services such as Jpush, PushY, etc are used there.

iOS, Android, SMS, Email: Users receive notifications on their devices.

Three problems are identified in this design:

- Single point of failure (SPOF): A single notification server means SPOF.

- Hard to scale: The notification system handles everything related to push notifications in one server. It is challenging to scale databases, caches, and different notification processing components independently.

- Performance bottleneck: Processing and sending notifications can be resource intensive. For example, constructing HTML pages and waiting for responses from third party services could take time. Handling everything in one system can result in the system overload, especially during peak hours.

High-level design (improved)

After enumerating challenges in the initial design, we improve the design as listed below:

- Move the database and cache out of the notification server.

- Add more notification servers and set up automatic horizontal scaling.

- Introduce message queues to decouple the system components.

Figure 10-10 shows the improved high-level design.

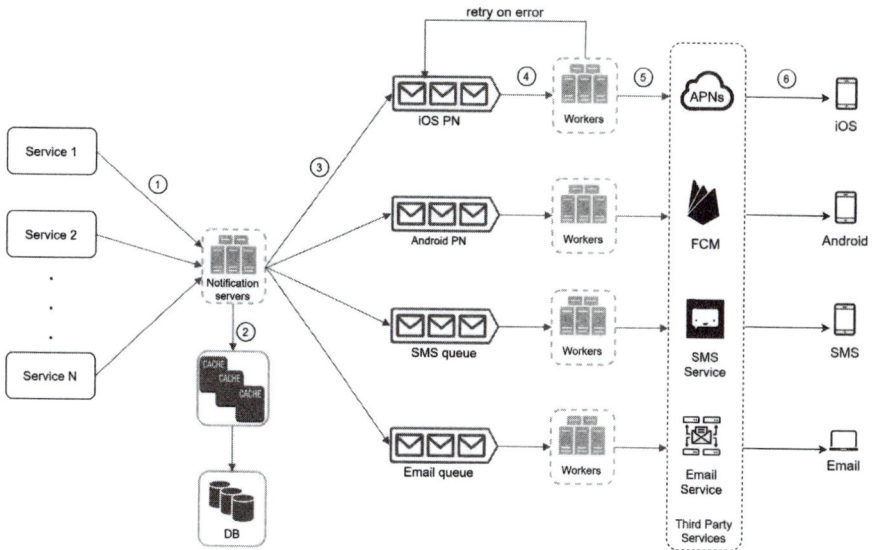

Figure 10-10

The best way to go through the above diagram is from left to right:

Service 1 to N: They represent different services that send notifications via APIs provided by notification servers.

Notification servers: They provide the following functionalities:

- Provide APIs for services to send notifications. Those APIs are only accessible internally or by verified clients to prevent spams.

- Carry out basic validations to verify emails, phone numbers, etc.

- Query the database or cache to fetch data needed to render a notification.

- Put notification data to message queues for parallel processing.

Here is an example of the API to send an email:

POST https://api.example.com/v/sms/send

Request body

```
{
  "to": [
    {
      "user_id": 123456
    }
  ],
  "from": {
    "email": "from_address@example.com"
  },
  "subject": "Hello, World!",
  "content": [
    {
      "type": "text/plain",
      "value": "Hello, World!"
    }
  ]
}
```

Cache: User info, device info, notification templates are cached.

DB: It stores data about user, notification, settings, etc.

Message queues: They remove dependencies between components. Message queues serve as buffers when high volumes of notifications are to be sent out. Each notification type is assigned with a distinct message queue so an outage in one third-party service will not affect other notification types.

Workers: Workers are a list of servers that pull notification events from message queues and send them to the corresponding third-party services.

Third-party services: Already explained in the initial design.

iOS, Android, SMS, Email: Already explained in the initial design.

Next, let us examine how every component works together to send a notification:

1. A service calls APIs provided by notification servers to send notifications.

2. Notification servers fetch metadata such as user info, device token, and notification setting from the cache or database.

3. A notification event is sent to the corresponding queue for processing. For instance, an iOS push notification event is sent to the iOS PN queue.

4. Workers pull notification events from message queues.

5. Workers send notifications to third party services.

6. Third-party services send notifications to user devices.

Step 3 - Design deep dive

In the high-level design, we discussed different types of notifications, contact info gathering flow, and notification sending/receiving flow. We will explore the following in deep dive:

- Reliability.

- Additional component and considerations: notification template, notification settings, rate limiting, retry mechanism, security in push notifications, monitor queued notifications and event tracking.

- Updated design.

Reliability

We must answer a few important reliability questions when designing a notification system in distributed environments.

How to prevent data loss?

One of the most important requirements in a notification system is that it cannot lose data. Notifications can usually be delayed or re-ordered, but never lost. To satisfy this requirement, the notification system persists notification data in a database and implements a retry mechanism. The notification log database is included for data persistence, as shown in Figure 10-11.

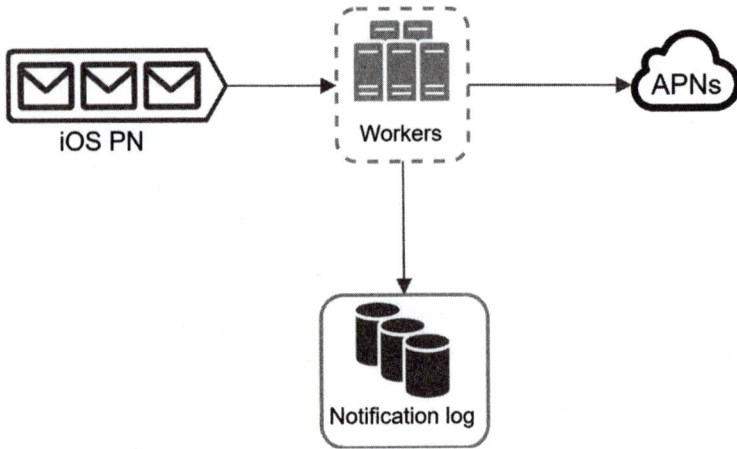

Figure 10-11

Will recipients receive a notification exactly once?

The short answer is no. Although notification is delivered exactly once most of the time, the distributed nature could result in duplicate notifications. To reduce the duplication occurrence, we introduce a dedupe mechanism and handle each failure case carefully. Here is a simple dedupe logic:

When a notification event first arrives, we check if it is seen before by checking the event ID. If it is seen before, it is discarded. Otherwise, we will send out the notification. For interested readers to explore why we cannot have exactly once delivery, refer to the reference material [5].

Additional components and considerations

We have discussed how to collect user contact info, send, and receive a notification. A notification system is a lot more than that. Here we discuss additional components including template reusing, notification settings, event tracking, system monitoring, rate limiting, etc.

Notification template

A large notification system sends out millions of notifications per day, and many of these notifications follow a similar format. Notification templates are introduced to avoid building every notification from scratch. A notification template is a preformatted notification to create your unique notification by customizing parameters, styling, tracking links, etc. Here is an example template of push notifications.

BODY:

You dreamed of it. We dared it. [ITEM NAME] is back — only until [DATE].

CTA:

Order Now. Or, Save My [ITEM NAME]

The benefits of using notification templates include maintaining a consistent format, reducing the margin error, and saving time.

Notification setting

Users generally receive way too many notifications daily and they can easily feel overwhelmed. Thus, many websites and apps give users fine-grained control over notification settings. This information is stored in the notification setting table, with the following fields:

user_id   bigInt

channel   varchar   # push notification, email or SMS

opt_in   boolean   # opt-in to receive notification

Before any notification is sent to a user, we first check if a user is opted-in to receive this type of notification.

Rate limiting

To avoid overwhelming users with too many notifications, we can limit the number of notifications a user can receive. This is important because receivers could turn off notifications completely if we send too often.

Retry mechanism

When a third-party service fails to send a notification, the notification will be added to the message queue for retrying. If the problem persists, an alert will be sent out to developers.

Security in push notifications

For iOS or Android apps, appKey and appSecret are used to secure push notification APIs [6]. Only authenticated or verified clients are allowed to send push notifications using our APIs. Interested users should refer to the reference material [6].

Monitor queued notifications

A key metric to monitor is the total number of queued notifications. If the number is large, the notification events are not processed fast enough by workers. To avoid delay in the notification delivery, more workers are needed. Figure 10-12 (credit to [7]) shows an example of queued messages to be processed.

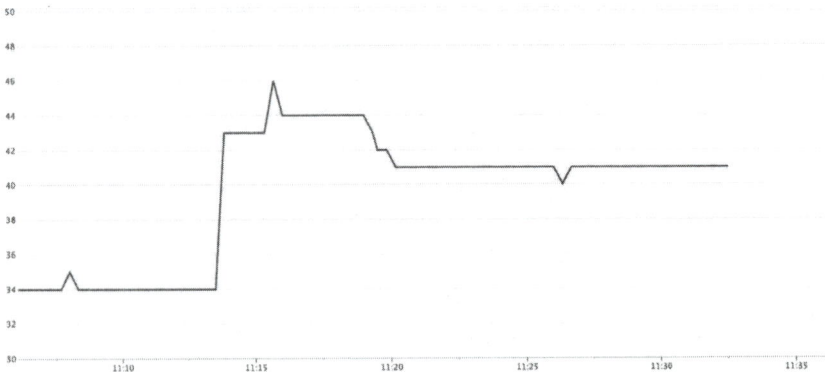

Figure 10-12

Events tracking

Notification metrics, such as open rate, click rate, and engagement are important in understanding customer behaviors. Analytics service implements events tracking. Integration between the notification system and the analytics service is usually required. Figure 10-13 shows an example of events that might be tracked for analytics purposes.

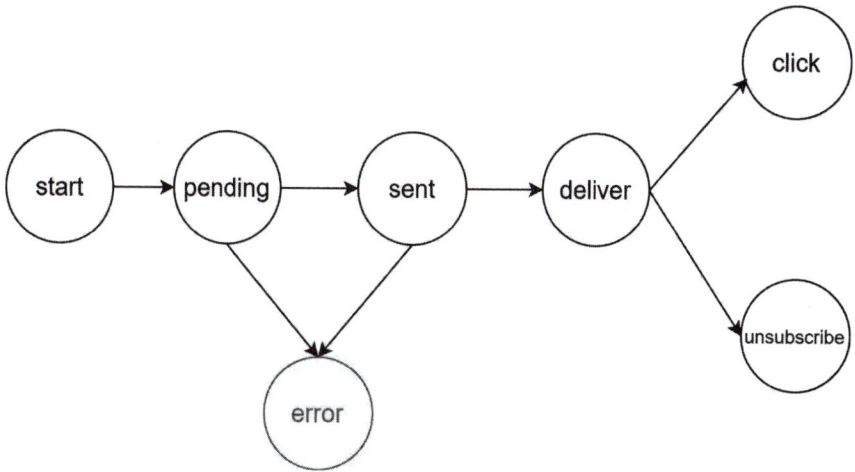

Figure 10-13

Updated design

Putting everything together, Figure 10-14 shows the updated notification system design.

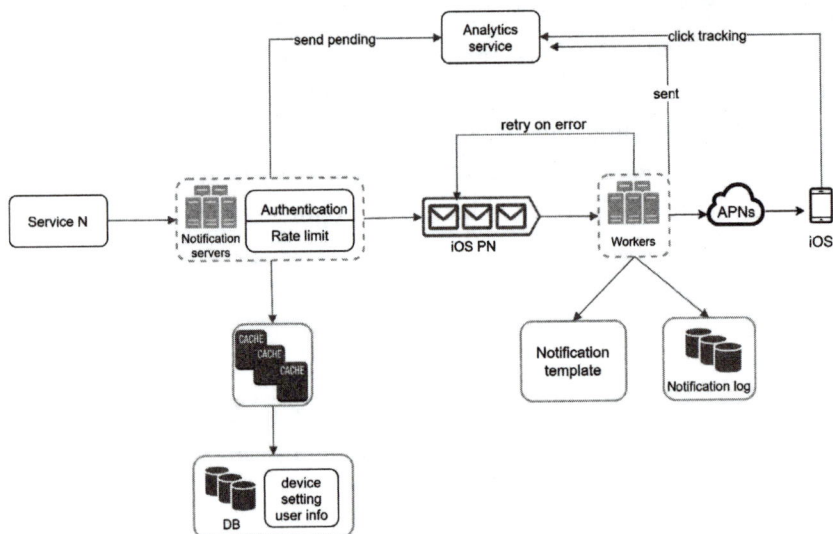

Figure 10-14

In this design, many new components are added in comparison with the previous design.

- The notification servers are equipped with two more critical features: authentication and rate-limiting.

- We also add a retry mechanism to handle notification failures. If the system fails to send notifications, they are put back in the messaging queue and the workers will retry for a predefined number of times.

- Furthermore, notification templates provide a consistent and efficient notification creation process.

- Finally, monitoring and tracking systems are added for system health checks and future improvements.

Step 4 - Wrap up

Notifications are indispensable because they keep us posted with important information. It could be a push notification about your favorite movie on Netflix, an email about discounts on new products, or a message about your online shopping payment confirmation.

In this chapter, we described the design of a scalable notification system that supports multiple notification formats: push notification, SMS message, and email. We adopted message queues to decouple system components.

Besides the high-level design, we dug deep into more components and optimizations.

- Reliability: We proposed a robust retry mechanism to minimize the failure rate.

- Security: AppKey/appSecret pair is used to ensure only verified clients can send notifications.

- Tracking and monitoring: These are implemented in any stage of a notification flow to capture important stats.

- Respect user settings: Users may opt-out of receiving notifications. Our system checks user settings first before sending notifications.

- Rate limiting: Users will appreciate a frequency capping on the number of notifications they receive.

Congratulations on getting this far! Now give yourself a pat on the back. Good job!

Reference materials

[1] Twilio SMS: https://www.twilio.com/sms

[2] Nexmo SMS: https://www.nexmo.com/products/sms

[3] Sendgrid: https://sendgrid.com/

[4] Mailchimp: https://mailchimp.com/

[5] You Cannot Have Exactly-Once Delivery:
https://bravenewgeek.com/you-cannot-have-exactly-once-delivery/

[6] Security in Push Notifications:
https://cloud.ibm.com/docs/services/mobilepush?topic=mobile-pushnotification-security-in-push-notifications

[7] Key metrics for RabbitMQ monitoring:
www.datadoghq.com/blog/rabbitmq-monitoring

DESIGN A NEWS FEED SYSTEM

In this chapter, you are asked to design a news feed system. What is news feed? According to the Facebook help page, "News feed is the constantly updating list of stories in the middle of your home page. News Feed includes status updates, photos, videos, links, app activity, and likes from people, pages, and groups that you follow on Facebook" [1]. This is a popular interview question. Similar questions commonly asked are to: design Facebook news feed, Instagram feed, Twitter timeline, etc.

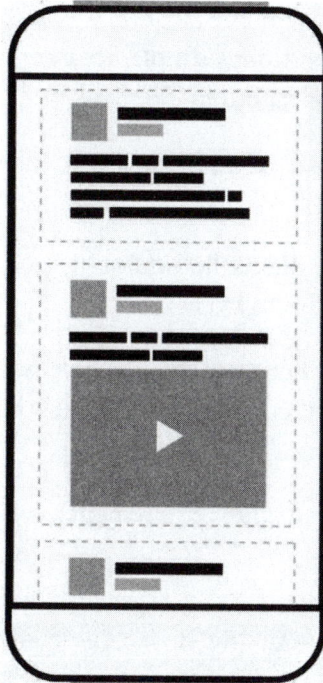

Figure 11-1

Step 1 - Understand the problem and establish design scope

The first set of clarification questions are to understand what the interviewer has in mind when she asks you to design a news feed system. At the very least, you should figure out what features to support. Here is an example of candidate-interviewer interaction:

Candidate: Is this a mobile app? Or a web app? Or both?
Interviewer: Both

Candidate: What are the important features?
Interview: A user can publish a post and see her friends' posts on the news feed page.

Candidate: Is the news feed sorted by reverse chronological order or any particular order such as topic scores? For instance, posts from your close friends have higher scores.
Interviewer: To keep things simple, let us assume the feed is sorted by reverse chronological order.

Candidate: How many friends can a user have?
Interviewer: 5000

Candidate: What is the traffic volume?
Interviewer: 10 million DAU

Candidate: Can feed contain images, videos, or just text?
Interviewer: It can contain media files, including both images and videos.

Now you have gathered the requirements, we focus on designing the system.

Step 2 - Propose high-level design and get buy-in

The design is divided into two flows: feed publishing and news feed building.

- Feed publishing: when a user publishes a post, corresponding data is written into cache and database. A post is populated to her friends' news feed.

- Newsfeed building: for simplicity, let us assume the news feed is built by aggregating friends' posts in reverse chronological order.

Newsfeed APIs

The news feed APIs are the primary ways for clients to communicate with servers. Those APIs are HTTP based that allow clients to perform actions, which include posting a status, retrieving news feed, adding friends, etc. We discuss two most important APIs: feed publishing API and news feed retrieval API.

Feed publishing API

To publish a post, a HTTP POST request will be sent to the server. The API is shown below:

POST /v1/me/feed

Params:

- content: content is the text of the post.

- auth_token: it is used to authenticate API requests.

Newsfeed retrieval API

The API to retrieve news feed is shown below:

GET /v1/me/feed

Params:

- auth_token: it is used to authenticate API requests.

Feed publishing

Figure 11-2 shows the high-level design of the feed publishing flow.

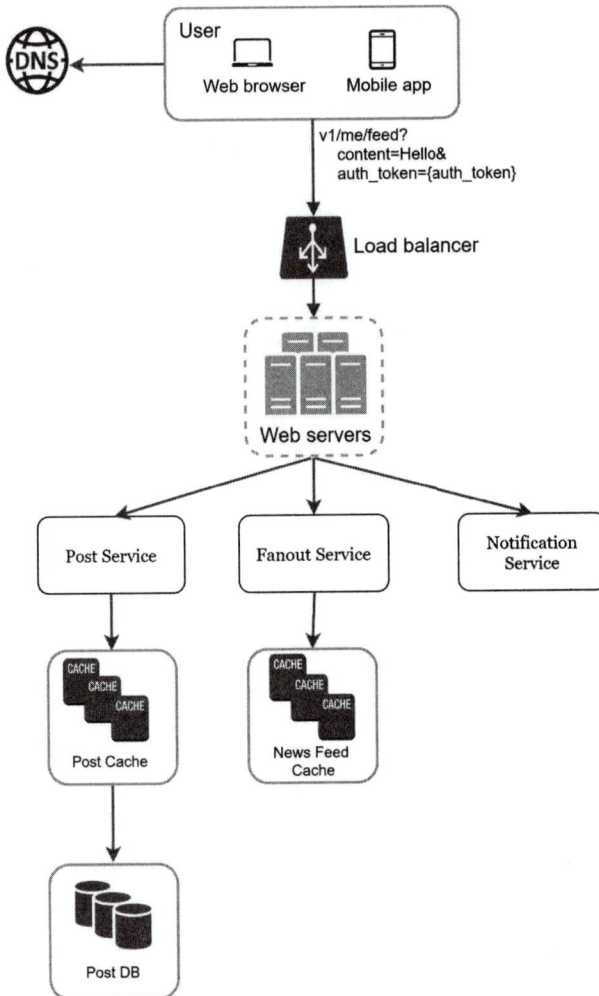

Figure 11-2

- User: a user can view news feeds on a browser or mobile app. A user makes a post with content "Hello" through API:

  /v1/me/feed?content=Hello&auth_token={auth_token}

- Load balancer: distribute traffic to web servers.

- Web servers: web servers redirect traffic to different internal services.

- Post service: persist post in the database and cache.

- Fanout service: push new content to friends' news feed. News-feed data is stored in the cache for fast retrieval.

- Notification service: inform friends that new content is available and send out push notifications.

Newsfeed building

In this section, we discuss how news feed is built behind the scenes. Figure 11-3 shows the high-level design:

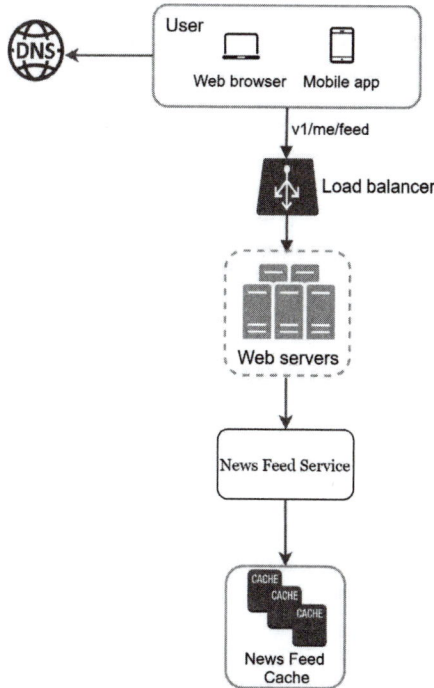

Figure 11-3

- User: a user sends a request to retrieve her news feed. The request looks like this: /v1/me/feed.

- Load balancer: load balancer redirects traffic to web servers.

- Web servers: web servers route requests to newsfeed service.

- Newsfeed service: news feed service fetches news feed from the cache.

- Newsfeed cache: store news feed IDs needed to render the news feed.

Step 3 - Design deep dive

The high-level design briefly covered two flows: feed publishing and news feed building. Here, we discuss those topics in more depth.

Feed publishing deep dive

Figure 11-4 outlines the detailed design for feed publishing. We have discussed most of components in high-level design, and we will focus on two components: web servers and fanout service.

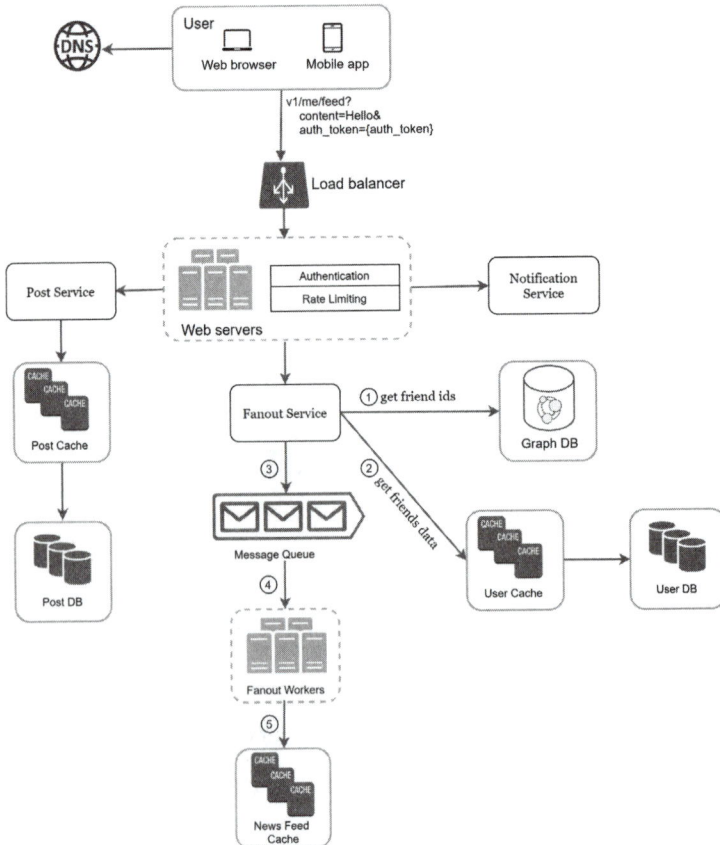

Figure 11-4

Web servers

Besides communicating with clients, web servers enforce authentication and rate-limiting. Only users signed in with valid *auth_token* are allowed to make posts. The system limits the number of posts a user can make within a certain period, vital to prevent spam and abusive content.

Fanout service

Fanout is the process of delivering a post to all friends. Two types of fanout models are: fanout on write (also called push model) and fanout on read (also called pull model). Both models have pros and cons. We explain their workflows and explore the best approach to support our system.

Fanout on write. With this approach, news feed is pre-computed during write time. A new post is delivered to friends' cache immediately after it is published.

Pros:

- The news feed is generated in real-time and can be pushed to friends immediately.

- Fetching news feed is fast because the news feed is pre-computed during write time.

Cons:

- If a user has many friends, fetching the friend list and generating news feeds for all of them are slow and time consuming. It is called hotkey problem.

- For inactive users or those rarely log in, pre-computing news feeds waste computing resources.

Fanout on read. The news feed is generated during read time. This is an on-demand model. Recent posts are pulled when a user loads her home page.

Pros:

- For inactive users or those who rarely log in, fanout on read works better because it will not waste computing resources on them.

- Data is not pushed to friends so there is no hotkey problem.

Cons:

- Fetching the news feed is slow as the news feed is not pre-computed.

We adopt a hybrid approach to get benefits of both approaches and avoid pitfalls in them. Since fetching the news feed fast is crucial, we use a push model for the majority of users. For celebrities or users who have many friends/followers, we let followers pull news content on-demand to avoid system overload. Consistent hashing is a useful technique to mitigate the hotkey problem as it helps to distribute requests/data more evenly.

Let us take a close look at the fanout service as shown in Figure 11-5.

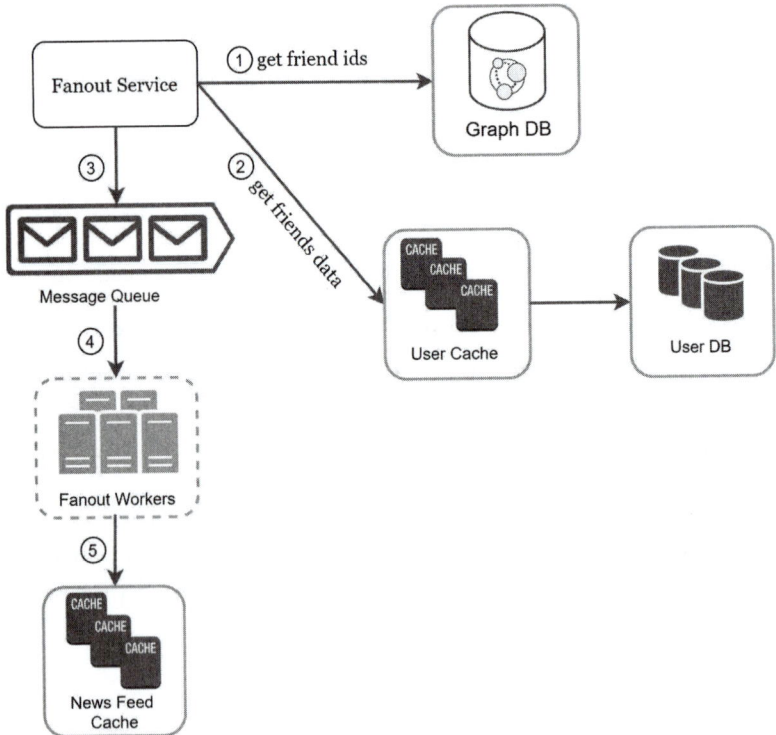

Figure 11-5

The fanout service works as follows:

1. Fetch friend IDs from the graph database. Graph databases are suited for managing friend relationship and friend recommendations. Interested readers wishing to learn more about this concept should refer to the reference material [2].

2. Get friends info from the user cache. The system then filters out friends based on user settings. For example, if you mute someone, her posts will not show up on your news feed even though you are still friends. Another reason why posts may not show is that a user could selectively share information with specific friends or hide it from other people.

3. Send friends list and new post ID to the message queue.

4. Fanout workers fetch data from the message queue and store news feed data in the news feed cache. You can think of the news feed cache as a *<post_id, user_id>* mapping table. Whenever a new post is made, it will be appended to the news feed table as shown in Figure 11-6. The memory consumption can become very large if we store the entire user and post objects in the cache. Thus, only IDs are stored. To keep the memory size small, we set a configurable limit. The chance of a user scrolling through thousands of posts in news feed is slim. Most users are only interested in the latest content, so the cache miss rate is low.

5. Store *<post_id, user_id >* in news feed cache. Figure 11-6 shows an example of what the news feed looks like in cache.

post_id	user_id
post_id	user_id
post_id	user_id
post_id	user_id
post_id	user_id
post_id	user_id
post_id	user_id
post_id	user_id

Figure 11-6

Newsfeed retrieval deep dive

Figure 11-7 illustrates the detailed design for news feed retrieval.

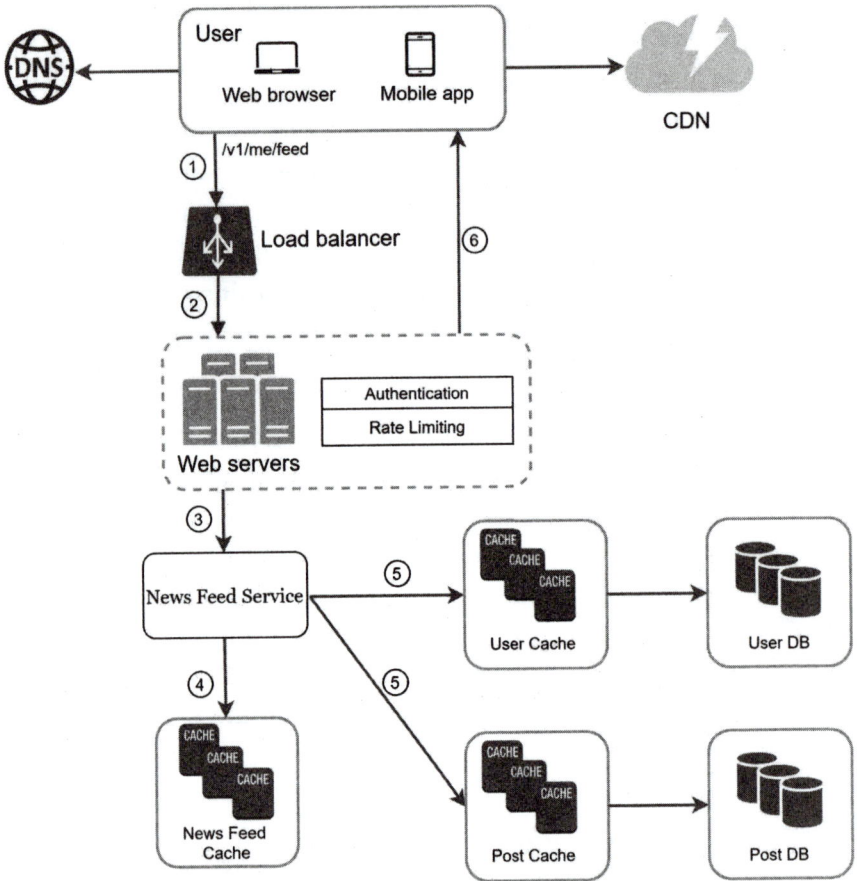

Figure 11-7

As shown in Figure 11-7, media content (images, videos, etc.) are stored in CDN for fast retrieval. Let us look at how a client retrieves news feed.

1. A user sends a request to retrieve her news feed. The request looks like this: */v1/me/feed*

2. The load balancer redistributes requests to web servers.

3. Web servers call the news feed service to fetch news feeds.

4. News feed service gets a list post IDs from the news feed cache.

5. A user's news feed is more than just a list of feed IDs. It contains username, profile picture, post content, post image, etc. Thus, the news feed service fetches the complete user and post objects from caches (user cache and post cache) to construct the fully hydrated news feed.

6. The fully hydrated news feed is returned in JSON format back to the client for rendering.

Cache architecture

Cache is extremely important for a news feed system. We divide the cache tier into 5 layers as shown in Figure 11-8.

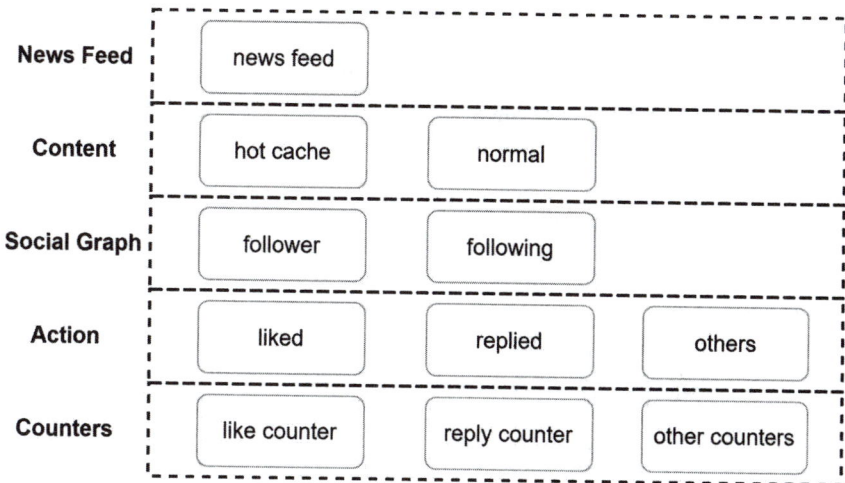

Figure 11-8

- News Feed: It stores IDs of news feeds.

- Content: It stores every post data. Popular content is stored in hot cache.

- Social Graph: It stores user relationship data.

- Action: It stores info about whether a user liked a post, replied a post, or took other actions on a post.

- Counters: It stores counters for like, reply, follower, following, etc.

Step 4 - Wrap up

In this chapter, we designed a news feed system. Our design contains two flows: feed publishing and news feed retrieval.

Like any system design interview questions, there is no perfect way to design a system. Every company has its unique constraints, and you must design a system to fit those constraints. Understanding the tradeoffs of your design and technology choices are important. If there are a few minutes left, you can talk about scalability issues. To avoid duplicated discussion, only high-level talking points are listed below.

Scaling the database:

- Vertical scaling vs Horizontal scaling
- SQL vs NoSQL
- Master-slave replication
- Read replicas
- Consistency models
- Database sharding

Other talking points:

- Keep web tier stateless
- Cache data as much as you can
- Support multiple data centers
- Lose couple components with message queues
- Monitor key metrics. For instance, QPS during peak hours and latency while users refreshing their news feed are interesting to monitor.

Congratulations on getting this far! Now give yourself a pat on the back. Good job!

Reference materials

[1] How News Feed Works:
https://www.facebook.com/help/327131014036297/

[2] Friend of Friend recommendations Neo4j and SQL Sever:
http://geekswithblogs.net/brendonpage/archive/2015/10/26/friend-of-
friend-recommendations-with-neo4j.aspx

12

DESIGN A CHAT SYSTEM

In this chapter we explore the design of a chat system. Almost everyone uses a chat app. Figure 12-1 shows some of the most popular apps in the marketplace.

Figure 12-1

A chat app performs different functions for different people. It is extremely important to nail down the exact requirements. For example, you do not want to design a system that focuses on group chat when the interviewer has one-on-one chat in mind. It is important to explore the feature requirements.

Step 1 - Understand the problem and establish design scope

It is vital to agree on the type of chat app to design. In the marketplace, there are one-on-one chat apps like Facebook Messenger, WeChat, and WhatsApp, office chat apps that focus on group chat like Slack, or game chat apps, like Discord, that focus on large group interaction and low voice chat latency.

The first set of clarification questions should nail down what the interviewer has in mind exactly when she asks you to design a chat system. At the very least, figure out if you should focus on a one-on-one chat or group chat app. Some questions you might ask are as follows:

Candidate: What kind of chat app shall we design? 1 on 1 or group based?
Interviewer: It should support both 1 on 1 and group chat.

Candidate: Is this a mobile app? Or a web app? Or both?
Interviewer: Both.

Candidate: What is the scale of this app? A startup app or massive scale?
Interviewer: It should support 50 million daily active users (DAU).

Candidate: For group chat, what is the group member limit?
Interviewer: A maximum of 100 people

Candidate: What features are important for the chat app? Can it support attachment?
Interviewer: 1 on 1 chat, group chat, online indicator. The system only supports text messages.

Candidate: Is there a message size limit?
Interviewer: Yes, text length should be less than 100,000 characters long.

Candidate: Is end-to-end encryption required?
Interviewer: Not required for now but we will discuss that if time allows.

Candidate: How long shall we store the chat history?
Interviewer: Forever.

In the chapter, we focus on designing a chat app like Facebook messenger, with an emphasis on the following features:

- A one-on-one chat with low delivery latency
- Small group chat (max of 100 people)

- Online presence

- Multiple device support. The same account can be logged in to multiple accounts at the same time.

- Push notifications

It is also important to agree on the design scale. We will design a system that supports 50 million DAU.

Step 2 - Propose high-level design and get buy-in

To develop a high-quality design, we should have a basic knowledge of how clients and servers communicate. In a chat system, clients can be either mobile applications or web applications. Clients do not communicate directly with each other. Instead, each client connects to a chat service, which supports all the features mentioned above. Let us focus on fundamental operations. The chat service must support the following functions:

- Receive messages from other clients.

- Find the right recipients for each message and relay the message to the recipients.

- If a recipient is not online, hold the messages for that recipient on the server until she is online.

Figure 12-2 shows the relationships between clients (sender and receiver) and the chat service.

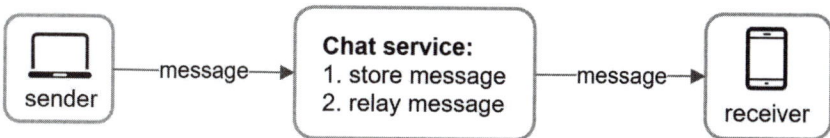

Figure 12-2

When a client intends to start a chat, it connects the chats service using one or more network protocols. For a chat service, the choice of network protocols is important. Let us discuss this with the interviewer.

Requests are initiated by the client for most client/server applications. This is also true for the sender side of a chat application. In Figure 12-2, when the sender sends a message to the receiver via the chat service, it uses the time-tested HTTP protocol, which is the most common web protocol. In this scenario, the client opens a HTTP connection with the chat service and sends the message, informing the service to send the message to the receiver. The keep-alive is efficient for this because the keep-alive header allows a client to maintain a persistent connection with the chat service. It also reduces the number of TCP handshakes. HTTP is a fine option on the sender side, and many popular chat applications such as Facebook [1] used HTTP initially to send messages.

However, the receiver side is a bit more complicated. Since HTTP is client-initiated, it is not trivial to send messages from the server. Over the years, many techniques are used to simulate a server-initiated connection: polling, long polling, and WebSocket. Those are important techniques widely used in system design interviews so let us examine each of them.

Polling

As shown in Figure 12-3, polling is a technique that the client periodically asks the server if there are messages available. Depending on polling frequency, polling could be costly. It could consume precious server resources to answer a question that offers no as an answer most of the time.

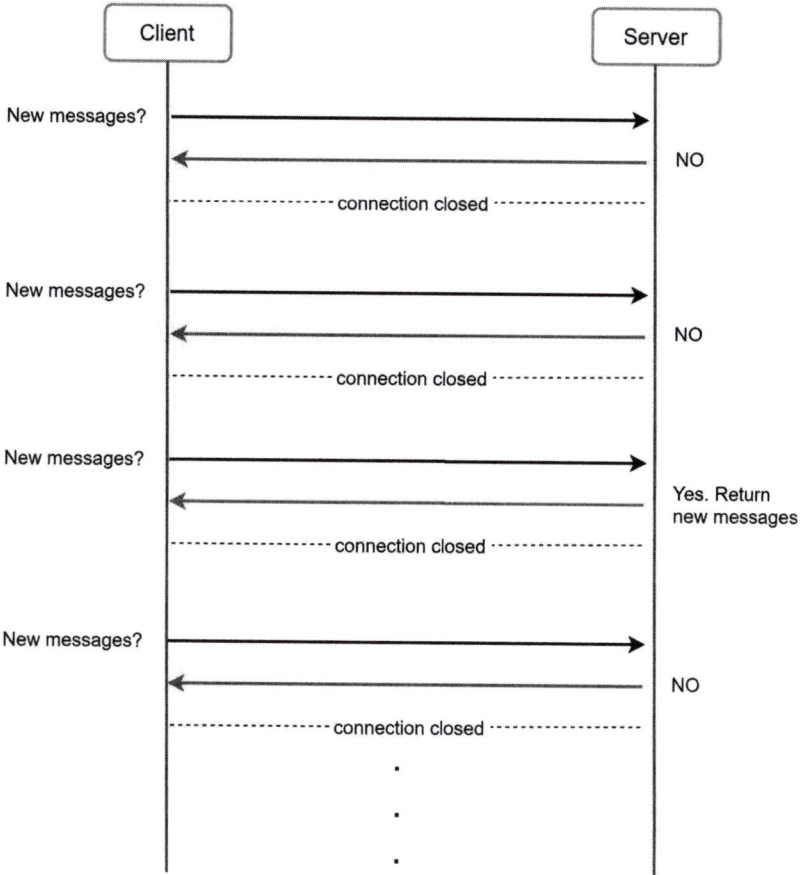

Figure 12-3

Long polling

Because polling could be inefficient, the next progression is long polling (Figure 12-4).

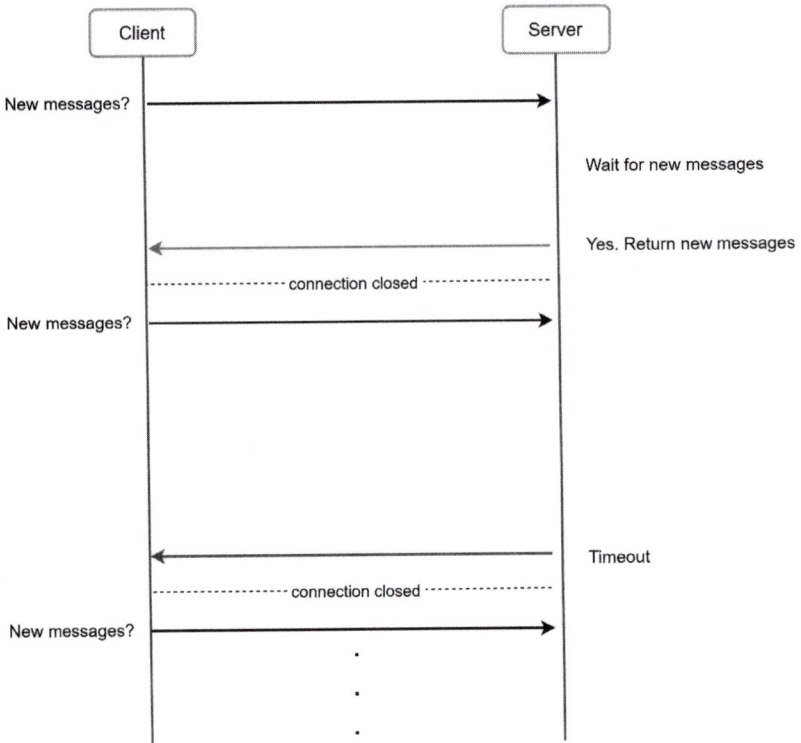

Figure 12-4

In long polling, a client holds the connection open until there are actually new messages available or a timeout threshold has been reached. Once the client receives new messages, it immediately sends another request to the server, restarting the process. Long polling has a few drawbacks:

- Sender and receiver may not connect to the same chat server. HTTP based servers are usually stateless. If you use round robin for load balancing, the server that receives the message might not have a long-polling connection with the client who receives the message.

- A server has no good way to tell if a client is disconnected.

- It is inefficient. If a user does not chat much, long polling still makes periodic connections after timeouts.

WebSocket

WebSocket is the most common solution for sending asynchronous updates from server to client. Figure 12-5 shows how it works.

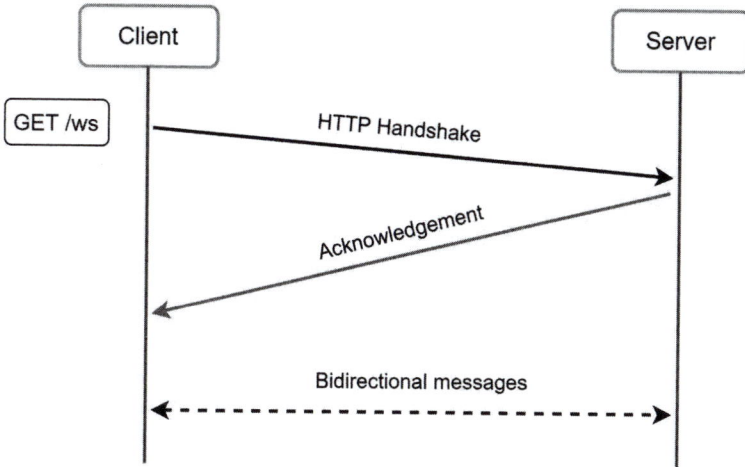

Figure 12-5

WebSocket connection is initiated by the client. It is bi-directional and persistent. It starts its life as a HTTP connection and could be "upgraded" via some well-defined handshake to a WebSocket connection. Through this persistent connection, a server could send updates to a client. WebSocket connections generally work even if a firewall is in place. This is because they use port 80 or 443 which are also used by HTTP/HTTPS connections.

Earlier we said that on the sender side HTTP is a fine protocol to use, but since WebSocket is bidirectional, there is no strong technical reason not to use it also for sending. Figure 12-6 shows how WebSockets (ws) is used for both sender and receiver sides.

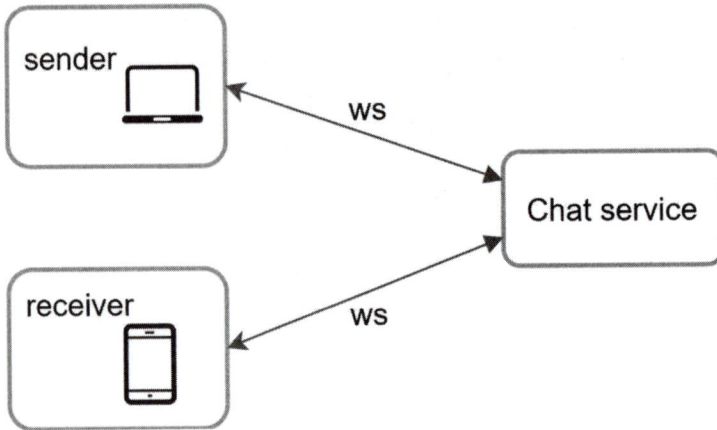

Figure 12-6

By using WebSocket for both sending and receiving, it simplifies the design and makes implementation on both client and server more straightforward. Since WebSocket connections are persistent, efficient connection management is critical on the server-side.

High-level design

Just now we mentioned that WebSocket was chosen as the main communication protocol between the client and server for its bidirectional communication, it is important to note that everything else does not have to be WebSocket. In fact, most features (sign up, login, user profile, etc) of a chat application could use the traditional request/response method over HTTP. Let us drill in a bit and look at the high-level components of the system.

As shown in Figure 12-7, the chat system is broken down into three major categories: stateless services, stateful services, and third-party integration.

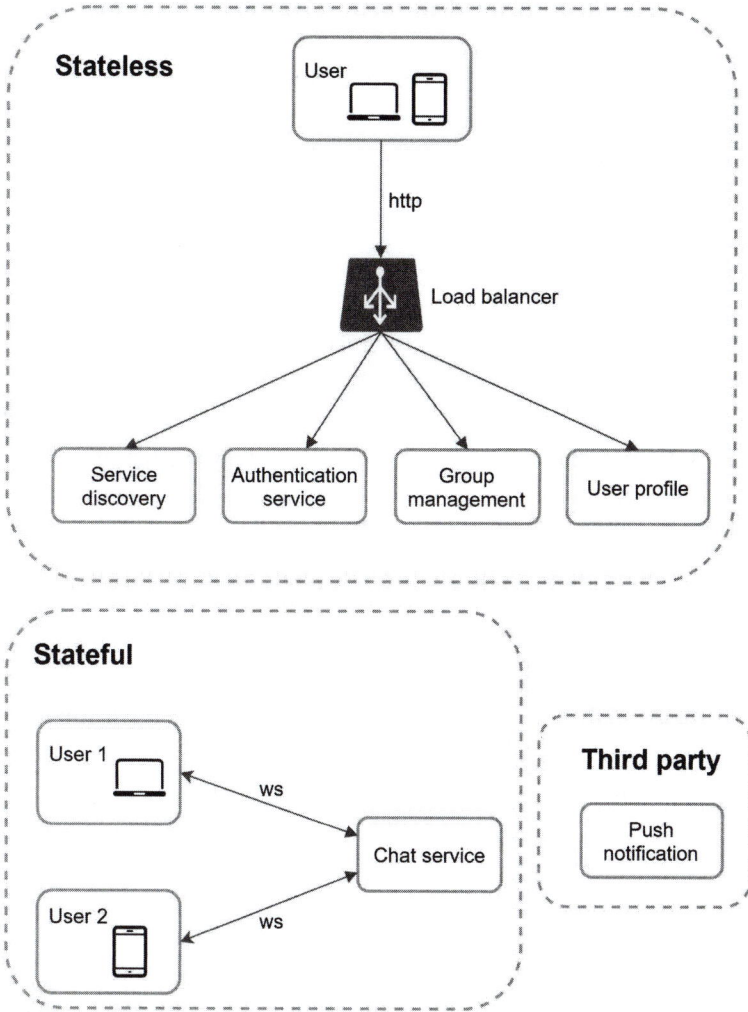

Figure 12-7

Stateless Services

Stateless services are traditional public-facing request/response services, used to manage the login, signup, user profile, etc. These are common features among many websites and apps.

Stateless services sit behind a load balancer whose job is to route requests to the correct services based on the request paths. These services can be monolithic or individual microservices. We do not need to build many of these stateless services by ourselves as there are services in the market that can be integrated easily. The one service that we will discuss more in deep dive is the service discovery. Its primary job is to give the client a list of DNS host names of chat servers that the client could connect to.

Stateful Service

The only stateful service is the chat service. The service is stateful because each client maintains a persistent network connection to a chat server. In this service, a client normally does not switch to another chat server as long as the server is still available. The service discovery coordinates closely with the chat service to avoid server overloading. We will go into detail in deep dive.

Third-party integration

For a chat app, push notification is the most important third-party integration. It is a way to inform users when new messages have arrived, even when the app is not running. Proper integration of push notification is crucial. Refer to Chapter 10 Design a notification system for more information.

Scalability

On a small scale, all services listed above could fit in one server. Even at the scale we design for, it is in theory possible to fit all user connections in one modern cloud server. The number of concurrent connections that a server can handle will most likely be the limiting factor. In our scenario, at 1M concurrent users, assuming each user connection needs 10K of memory on the server (this is a very rough figure and very dependent on the language choice), it only needs about 10GB of memory to hold all the connections on one box.

If we propose a design where everything fits in one server, this may raise a big red flag in the interviewer's mind. No technologist would design such a scale in a single server. Single server design is a deal breaker due to many factors. The single point of failure is the biggest among them.

However, it is perfectly fine to start with a single server design. Just make sure the interviewer knows this is a starting point. Putting everything we mentioned together, Figure 12-8 shows the adjusted high-level design.

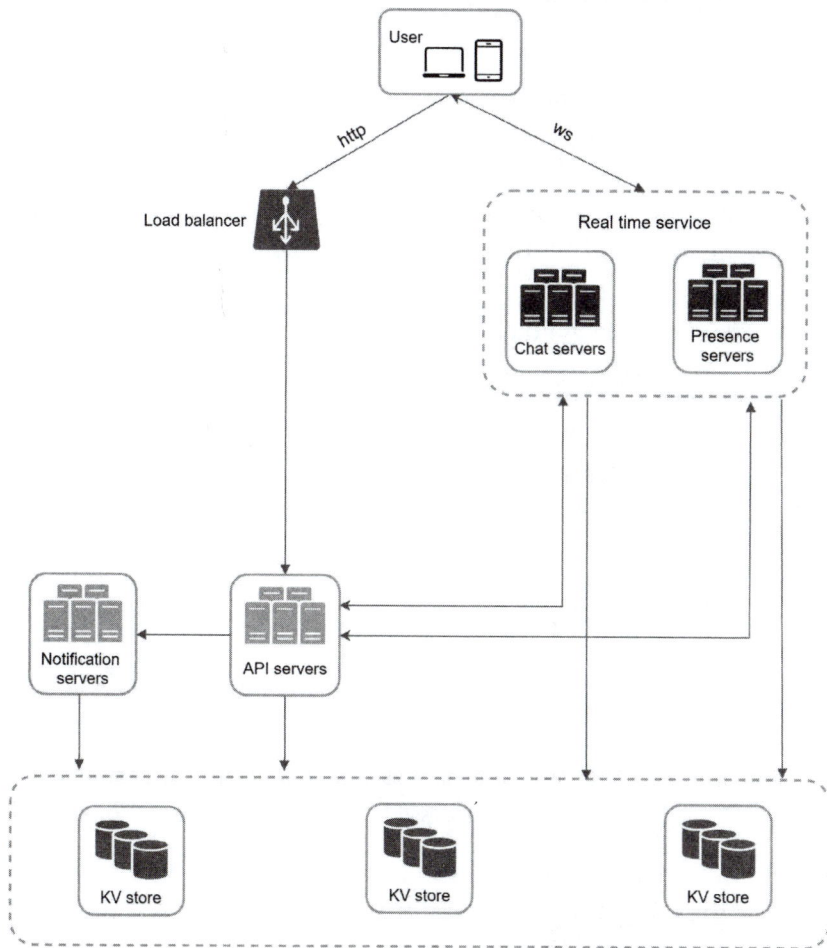

Figure 12-8

In Figure 12-8, the client maintains a persistent WebSocket connection to a chat server for real-time messaging.

- Chat servers facilitate message sending/receiving.

- Presence servers manage online/offline status.

- API servers handle everything including user login, signup, change profile, etc.

- Notification servers send push notifications.

- Finally, the key-value store is used to store chat history. When an offline user comes online, she will see all her previous chat history.

Storage

At this point, we have servers ready, services up running and third-party integrations complete. Deep down the technical stack is the data layer. Data layer usually requires some effort to get it correct. An important decision we must make is to decide on the right type of database to use: relational databases or NoSQL databases? To make an informed decision, we will examine the data types and read/write patterns.

Two types of data exist in a typical chat system. The first is generic data, such as user profile, setting, user friends list. These data are stored in robust and reliable relational databases. Replication and sharding are common techniques to satisfy availability and scalability requirements.

The second is unique to chat systems: chat history data. It is important to understand the read/write pattern.

- The amount of data is enormous for chat systems. A previous study [2] reveals that Facebook messenger and Whatsapp process 60 billion messages a day.

- Only recent chats are accessed frequently. Users do not usually look up for old chats.

- Although very recent chat history is viewed in most cases, users might use features that require random access of data, such as search, view your mentions, jump to specific messages, etc. These cases should be supported by the data access layer.

- The read to write ratio is about 1:1 for 1 on 1 chat apps.

Selecting the correct storage system that supports all of our use cases is crucial. We recommend key-value stores for the following reasons:

- Key-value stores allow easy horizontal scaling.

- Key-value stores provide very low latency to access data.

- Relational databases do not handle long tail [3] of data well. When the indexes grow large, random access is expensive.

- Key-value stores are adopted by other proven reliable chat applications. For example, both Facebook messenger and Discord use key-value stores. Facebook messenger uses HBase [4], and Discord uses Cassandra [5].

Data models

Just now, we talked about using key-value stores as our storage layer. The most important data is message data. Let us take a close look.

Message table for 1 on 1 chat

Figure 12-9 shows the message table for 1 on 1 chat. The primary key is *message_id*, which helps to decide message sequence. We cannot rely on *created_at* to decide the message sequence because two messages can be created at the same time.

message	
message_id	**bigint**
message_from	bigint
message_to	bitint
content	text
created_at	timestamp

Figure 12-9

Message table for group chat

Figure 12-10 shows the message table for group chat. The composite primary key is *(channel_id, message_id)*. Channel and group represent the same meaning here. *channel_id* is the partition key because all queries in a group chat operate in a channel.

group_message	
channel_id	bigint
message_id	bigint
user_id	bigint
content	text
created_at	timestamp

Figure 12-10

Message ID

How to generate *message_id* is an interesting topic worth exploring. *Message_id* carries the responsibility of ensuring the order of messages. To ascertain the order of messages, *message_id* must satisfy the following two requirements:

- IDs must be unique.
- IDs should be sortable by time, meaning new rows have higher IDs than old ones.

How can we achieve those two guarantees? The first idea that comes to mind is the "*auto_increment*" keyword in MySql. However, NoSQL databases usually do not provide such a feature.

The second approach is to use a global 64-bit sequence number generator like Snowflake [6]. This is discussed in "Chapter 7: Design a unique ID generator in a distributed system".

The final approach is to use local sequence number generator. Local means IDs are only unique within a group. The reason why local IDs work is that maintaining message sequence within one-on-one channel or a group channel is sufficient. This approach is easier to implement in comparison to the global ID implementation.

Step 3 - Design deep dive

In a system design interview, usually you are expected to dive deep into some of the components in the high-level design. For the chat system, service discovery, messaging flows, and online/offline indicators worth deeper exploration.

Service discovery

The primary role of service discovery is to recommend the best chat server for a client based on the criteria like geographical location, server capacity, etc. Apache Zookeeper [7] is a popular open-source solution for service discovery. It registers all the available chat servers and picks the best chat server for a client based on predefined criteria.

Figure 12-11 shows how service discovery (Zookeeper) works.

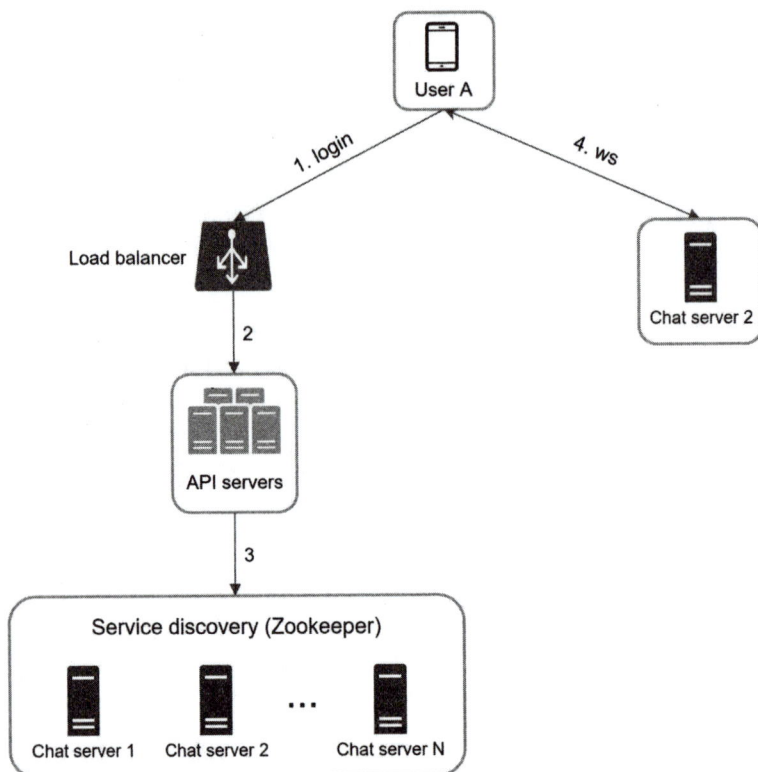

Figure 12-11

1. User A tries to log in to the app.

2. The load balancer sends the login request to API servers.

3. After the backend authenticates the user, service discovery finds the best chat server for User A. In this example, server 2 is chosen and the server info is returned back to User A.

4. User A connects to chat server 2 through WebSocket.

Message flows

It is interesting to understand the end-to-end flow of a chat system. In this section, we will explore 1 on 1 chat flow, message synchronization across multiple devices and group chat flow.

I on I chat flow

Figure 12-12 explains what happens when User A sends a message to User B.

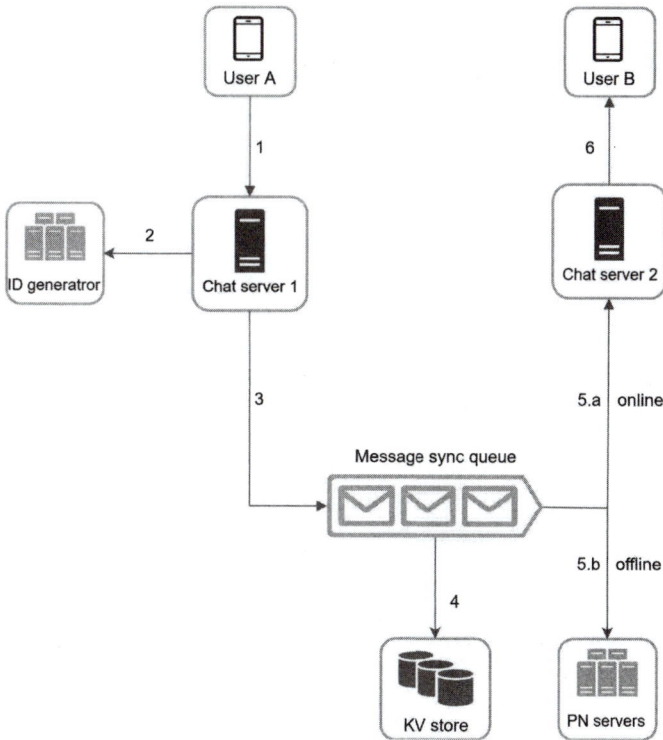

Figure 12-12

1. User A sends a chat message to Chat server 1.

2. Chat server 1 obtains a message ID from the ID generator.

3. Chat server 1 sends the message to the message sync queue.

4. The message is stored in a key-value store.

5. a. If User B is online, the message is forwarded to Chat server 2 where User B is connected.

   b. If User B is offline, a push notification is sent from push notification (PN) servers.

6. Chat server 2 forwards the message to User B. There is a persistent WebSocket connection between User B and Chat server 2.

Message synchronization across multiple devices

Many users have multiple devices. We will explain how to sync messages across multiple devices. Figure 12-13 shows an example of message synchronization.

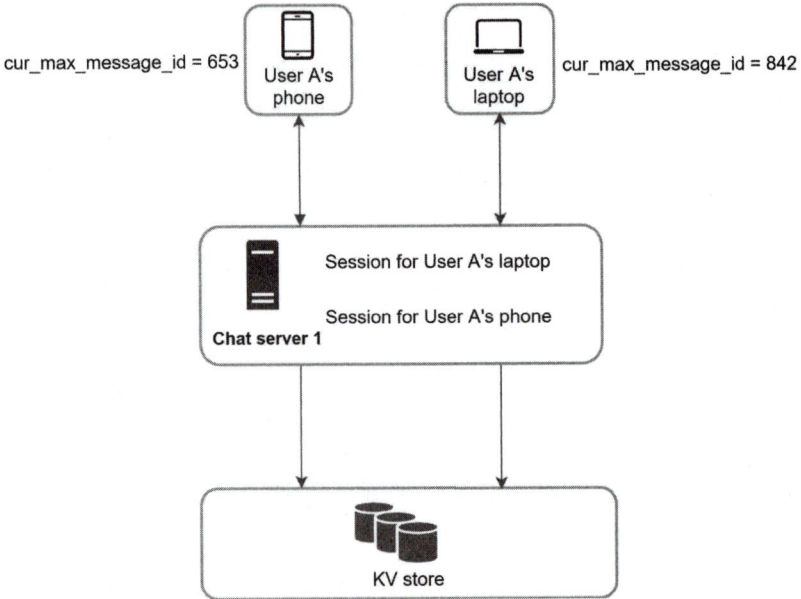

Figure 12-13

In Figure 12-13, user A has two devices: a phone and a laptop. When User A logs in to the chat app with her phone, it establishes a WebSocket connection with Chat server 1. Similarly, there is a connection between the laptop and Chat server 1.

Each device maintains a variable called *cur_max_message_id*, which keeps track of the latest message ID on the device. Messages that satisfy the following two conditions are considered as news messages:

- The recipient ID is equal to the currently logged-in user ID.
- Message ID in the key-value store is larger than *cur_max_message_id*.

With distinct *cur_max_message_id* on each device, message synchronization is easy as each device can get new messages from the KV store.

Small group chat flow

In comparison to the one-on-one chat, the logic of group chat is more complicated. Figures 12-14 and 12-15 explain the flow.

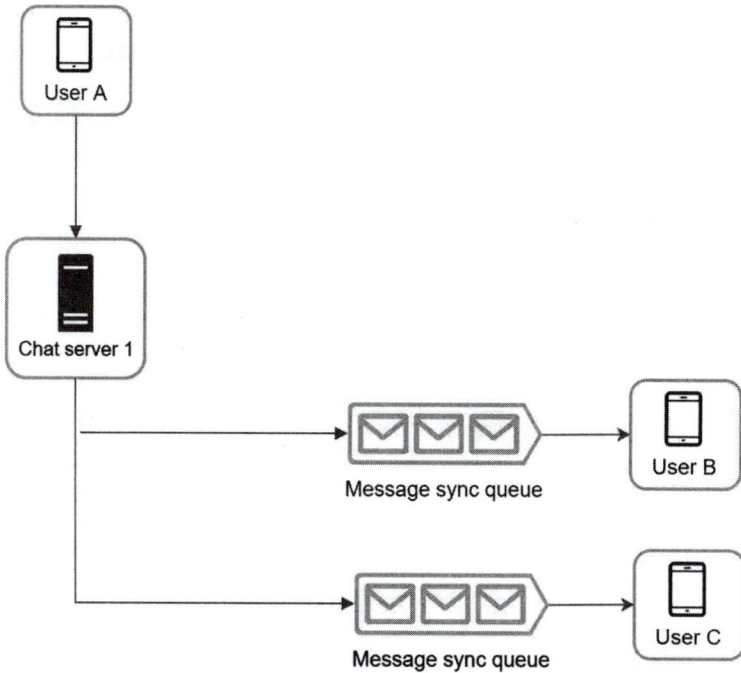

Figure 12-14

Figure 12-14 explains what happens when User A sends a message in a group chat. Assume there are 3 members in the group (User A, User B and user C). First, the message from User A is copied to each group member's message sync queue: one for User B and the second for User C. You can think of the message sync queue as an inbox for a recipient. This design choice is good for small group chat because:

- it simplifies message sync flow as each client only needs to check its own inbox to get new messages.

- when the group number is small, storing a copy in each recipient's inbox is not too expensive.

WeChat uses a similar approach, and it limits a group to 500 members

[8]. However, for groups with a lot of users, storing a message copy for each member is not acceptable.

On the recipient side, a recipient can receive messages from multiple users. Each recipient has an inbox (message sync queue) which contains messages from different senders. Figure 12-15 illustrates the design.

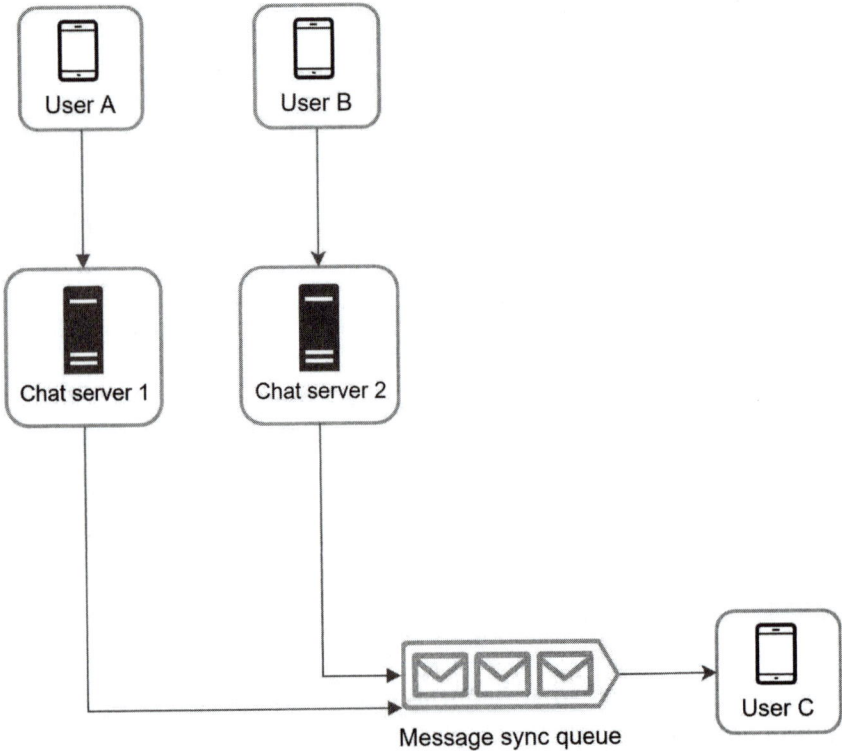

Figure 12-15

Online presence

An online presence indicator is an essential feature of many chat applications. Usually, you can see a green dot next to a user's profile picture or username. This section explains what happens behind the scenes.

In the high-level design, presence servers are responsible for managing online status and communicating with clients through WebSocket.

There are a few flows that will trigger online status change. Let us examine each of them.

User login

The user login flow is explained in the "Service Discovery" section. After a WebSocket connection is built between the client and the real-time service, user A's online status and *last_active_at* timestamp are saved in the KV store. Presence indicator shows the user is online after she logs in.

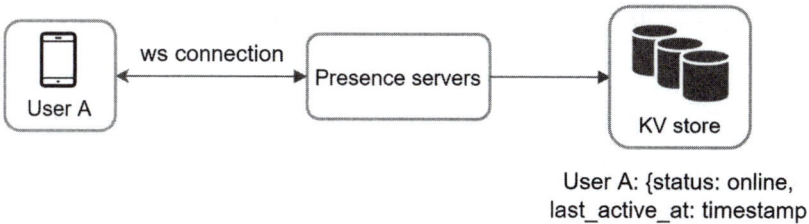

User A: {status: online,
last_active_at: timestamp}

Figure 12-16

User logout

When a user logs out, it goes through the user logout flow as shown in Figure 12-17. The online status is changed to offline in the KV store. The presence indicator shows a user is offline.

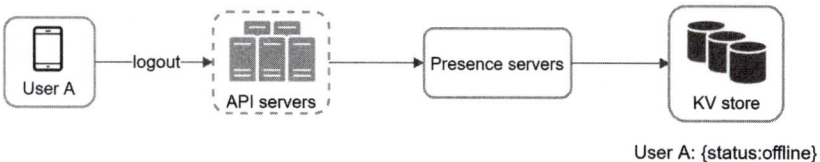

User A: {status:offline}

Figure 12-17

User disconnection

We all wish our internet connection is consistent and reliable. However, that is not always the case; thus, we must address this issue in our design. When a user disconnects from the internet, the persistent connection between the client and server is lost. A naive way to handle user disconnection is to mark the user as offline and change the status to online when the connection re-establishes. However, this approach has a major flaw. It is common for users to disconnect and reconnect to the internet frequently in a short time. For example, network connections can be on and off while a user goes through a tunnel. Updating online status on every disconnect/reconnect would make the presence indicator change too often, resulting in poor user experience.

We introduce a heartbeat mechanism to solve this problem. Periodically, an online client sends a heartbeat event to presence servers. If presence servers receive a heartbeat event within a certain time, say x seconds from the client, a user is considered as online. Otherwise, it is offline.

In Figure 12-18, the client sends a heartbeat event to the server every 5 seconds. After sending 3 heartbeat events, the client is disconnected and does not reconnect within x = 30 seconds (This number is arbitrarily chosen to demonstrate the logic). The online status is changed to offline.

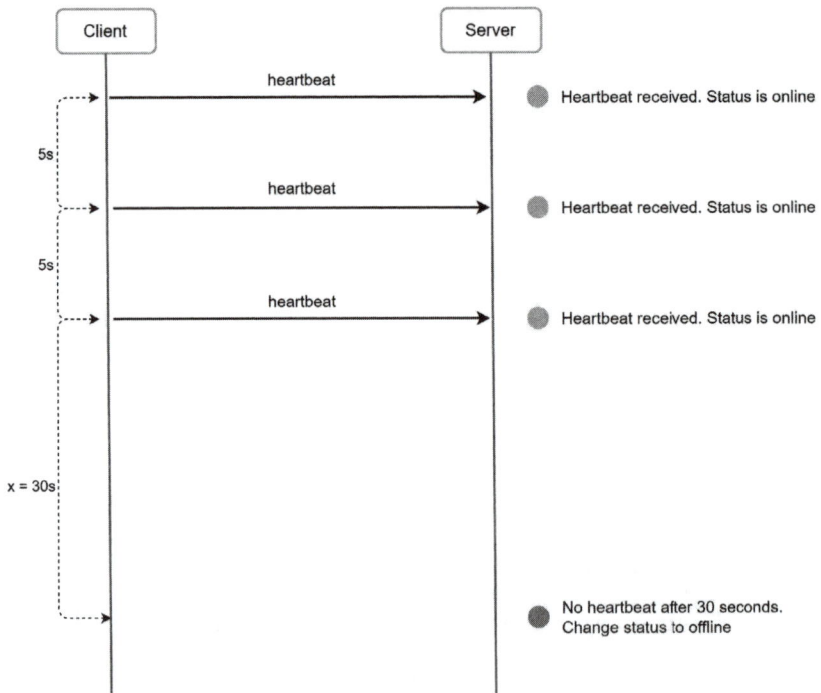

Figure 12-18

Online status fanout

How do user A's friends know about the status changes? Figure 12-19 explains how it works. Presence servers use a publish-subscribe model, in which each friend pair maintains a channel. When User A's online status changes, it publishes the event to three channels, channel A-B, A-C, and A-D. Those three channels are subscribed by User B, C, and D, respectively. Thus, it is easy for friends to get online status updates. The communication between clients and servers is through real-time WebSocket.

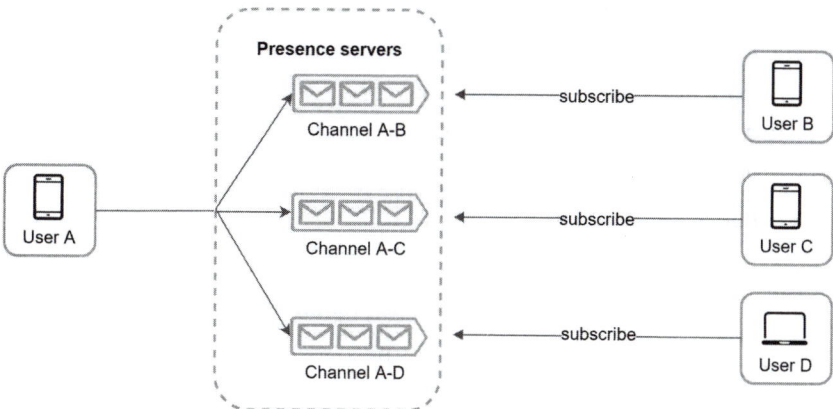

Figure 12-19

The above design is effective for a small user group. For instance, WeChat uses a similar approach because its user group is capped to 500. For larger groups, informing all members about online status is expensive and time consuming. Assume a group has 100,000 members. Each status change will generate 100,000 events. To solve the performance bottleneck, a possible solution is to fetch online status only when a user enters a group or manually refreshes the friend list.

Step 4 - Wrap up

In this chapter, we presented a chat system architecture that supports both 1-to-1 chat and small group chat. WebSocket is used for real-time communication between the client and server. The chat system contains the following components: chat servers for real-time messaging, presence servers for managing online presence, push notification servers for sending push notifications, key-value stores for chat history persistence and API servers for other functionalities.

If you have extra time at the end of the interview, here are additional talking points:

- Extend the chat app to support media files such as photos and videos. Media files are significantly larger than text in size. Compression, cloud storage, and thumbnails are interesting topics to talk about.

- End-to-end encryption. Whatsapp supports end-to-end encryption for messages. Only the sender and the recipient can read messages. Interested readers should refer to the article in the reference materials [9].

- Caching messages on the client-side is effective to reduce the data transfer between the client and server.

- Improve load time. Slack built a geographically distributed network to cache users' data, channels, etc. for better load time [10].

- Error handling.

  o The chat server error. There might be hundreds of thousands, or even more persistent connections to a chat server. If a chat server goes offline, service discovery (Zookeeper) will provide a new chat server for clients to establish new connections with.

  o Message resent mechanism. Retry and queueing are common techniques for resending messages.

Congratulations on getting this far! Now give yourself a pat on the back. Good job!

Reference materials

[1] Erlang at Facebook:
https://www.erlang-factory.com/upload/presentations/31/
EugeneLetuchy-ErlangatFacebook.pdf

[2] Messenger and WhatsApp process 60 billion messages a day:
https://www.theverge.com/2016/4/12/11415198/facebook-messenger-
whatsapp-number-messages-vs-sms-f8-2016

[3] Long tail: https://en.wikipedia.org/wiki/Long_tail

[4] The Underlying Technology of Messages:
https://www.facebook.com/notes/facebook-engineering/the-underlying-
technology-of-messages/454991608919/

[5] How Discord Stores Billions of Messages:
https://blog.discordapp.com/how-discord-stores-billions-of-messages-
7fa6ec7ee4c7

[6] Announcing Snowflake:
https://blog.twitter.com/engineering/en_us/a/2010/announcing-
snowflake.html

[7] Apache ZooKeeper: https://zookeeper.apache.org/

[8] From nothing: the evolution of WeChat background system (Article
in Chinese):
https://www.infoq.cn/article/the-road-of-the-growth-
weixin-background

[9] End-to-end encryption:
https://faq.whatsapp.com/en/android/28030015/

[10] Flannel: An Application-Level Edge Cache to Make Slack Scale:
https://slack.engineering/flannel-an-application-level-edge-cache-to-
make-slack-scale-b8a6400e2f6b

13

DESIGN A SEARCH AUTOCOMPLETE SYSTEM

When searching on Google or shopping at Amazon, as you type in the search box, one or more matches for the search term are presented to you. This feature is referred to as autocomplete, typeahead, search-as-you-type, or incremental search. Figure 13-1 presents an example of a Google search showing a list of autocompleted results when "dinner" is typed into the search box. Search autocomplete is an important feature of many products. This leads us to the interview question: design a search autocomplete system, also called "design top k" or "design top k most searched queries".

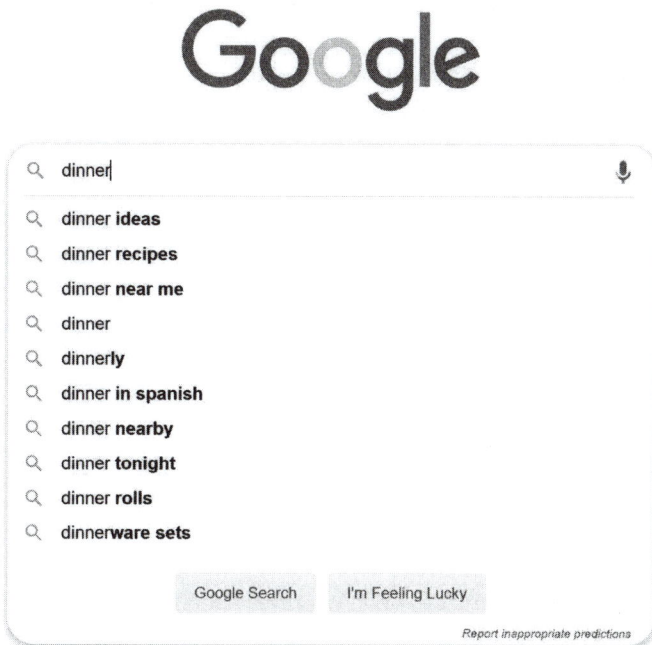

Figure 13-1

Step 1 - Understand the problem and establish design scope

The first step to tackle any system design interview question is to ask enough questions to clarify requirements. Here is an example of candidate-interviewer interaction:

> **Candidate**: Is the matching only supported at the beginning of a search query or in the middle as well?
> **Interviewer**: Only at the beginning of a search query.

> **Candidate**: How many autocomplete suggestions should the system return?
> **Interviewer**: 5

> **Candidate**: How does the system know which 5 suggestions to return?
> **Interviewer**: This is determined by popularity, decided by the historical query frequency.

> **Candidate**: Does the system support spell check?
> **Interviewer**: No, spell check or autocorrect is not supported.

> **Candidate**: Are search queries in English?
> **Interviewer**: Yes. If time allows at the end, we can discuss multi-language support.

> **Candidate**: Do we allow capitalization and special characters?
> **Interviewer**: No, we assume all search queries have lowercase alphabetic characters.

> **Candidate**: How many users use the product?
> **Interviewer**: 10 million DAU.

Requirements

Here is a summary of the requirements:

- Fast response time: As a user types a search query, autocomplete suggestions must show up fast enough. An article about Facebook's autocomplete system [1] reveals that the system needs to return results within 100 milliseconds. Otherwise it will cause stuttering.

- Relevant: Autocomplete suggestions should be relevant to the search term.

- Sorted: Results returned by the system must be sorted by popularity or other ranking models.

- Scalable: The system can handle high traffic volume.

- Highly available: The system should remain available and accessible when part of the system is offline, slows down, or experiences unexpected network errors.

Back of the envelope estimation

- Assume 10 million daily active users (DAU).

- An average person performs 10 searches per day.

- 20 bytes of data per query string:

  o   Assume we use ASCII character encoding. 1 character = 1 byte

  o   Assume a query contains 4 words, and each word contains 5 characters on average.

  o   That is 4 x 5 = 20 bytes per query.

- For every character entered into the search box, a client sends a request to the backend for autocomplete suggestions. On average, 20 requests are sent for each search query. For example, the following 6 requests are sent to the backend by the time you finish typing "dinner".

search?q=d

search?q=di

search?q=din

search?q=dinn

search?q=dinne

search?q=dinner

- ~24,000 query per second (QPS) = 10,000,000 users * 10 queries / day * 20 characters / 24 hours / 3600 seconds.

- Peak QPS = QPS * 2 = ~48,000

- Assume 20% of the daily queries are new. 10 million * 10 queries / day * 20 byte per query * 20% = 0.4 GB. This means 0.4GB of new data is added to storage daily.

Step 2 - Propose high-level design and get buy-in

At the high-level, the system is broken down into two:

- Data gathering service: It gathers user input queries and aggregates them in real-time. Real-time processing is not practical for large data sets; however, it is a good starting point. We will explore a more realistic solution in deep dive.

- Query service: Given a search query or prefix, return 5 most frequently searched terms.

Data gathering service

Let us use a simplified example to see how data gathering service works. Assume we have a frequency table that stores the query string and its frequency as shown in Figure 13-2. In the beginning, the frequency table is empty. Later, users enter queries "twitch", "twitter", "twitter," and "twillo" sequentially. Figure 13-2 shows how the frequency table is updated.

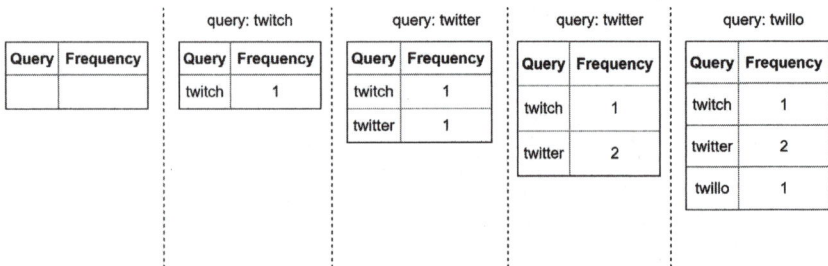

Figure 13-2

Query service

Assume we have a frequency table as shown in Table 13-1. It has two fields.

- Query: it stores the query string.

- Frequency: it represents the number of times a query has been searched.

Query	Frequency
twitter	35
twitch	29
twilight	25
twin peak	21
twitch prime	18
twitter search	14
twillo	10
twin peak sf	8

Table 13-1

When a user types "tw" in the search box, the following top 5 searched queries are displayed (Figure 13-3), assuming the frequency table is based on Table 13-1.

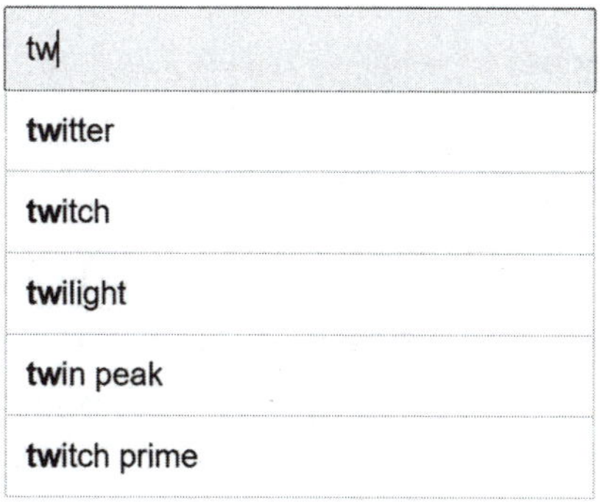

Figure 13-3

To get top 5 frequently searched queries, execute the following SQL query:

```
SELECT * FROM frequency_table
WHERE query Like `prefix%`
ORDER BY frequency DESC
LIMIT 5
```

Figure 13-4

This is an acceptable solution when the data set is small. When it is large, accessing the database becomes a bottleneck. We will explore optimizations in deep dive.

Step 3 - Design deep dive

In the high-level design, we discussed data gathering service and query service. The high-level design is not optimal, but it serves as a good starting point. In this section, we will dive deep into a few components and explore optimizations as follows:

- Trie data structure
- Data gathering service
- Query service
- Scale the storage
- Trie operations

Trie data structure

Relational databases are used for storage in the high-level design. However, fetching the top 5 search queries from a relational database is inefficient. The data structure trie (prefix tree) is used to overcome the problem. As trie data structure is crucial for the system, we will dedicate significant time to design a customized trie. Please note that some of the ideas are from articles [2] and [3].

Understanding the basic trie data structure is essential for this interview question. However, this is more of a data structure question than a system design question. Besides, many online materials explain this concept. In this chapter, we will only discuss an overview of the trie data structure and focus on how to optimize the basic trie to improve response time.

Trie (pronounced "try") is a tree-like data structure that can compactly store strings. The name comes from the word retrieval, which indicates it is designed for string retrieval operations. The main idea of trie consists of the following:

- A trie is a tree-like data structure.

- The root represents an empty string.

- Each node stores a character and has 26 children, one for each possible character. To save space, we do not draw empty links.

- Each tree node represents a single word or a prefix string.

Figure 13-5 shows a trie with search queries "tree", "try", "true", "toy", "wish", "win". Search queries are highlighted with a thicker border.

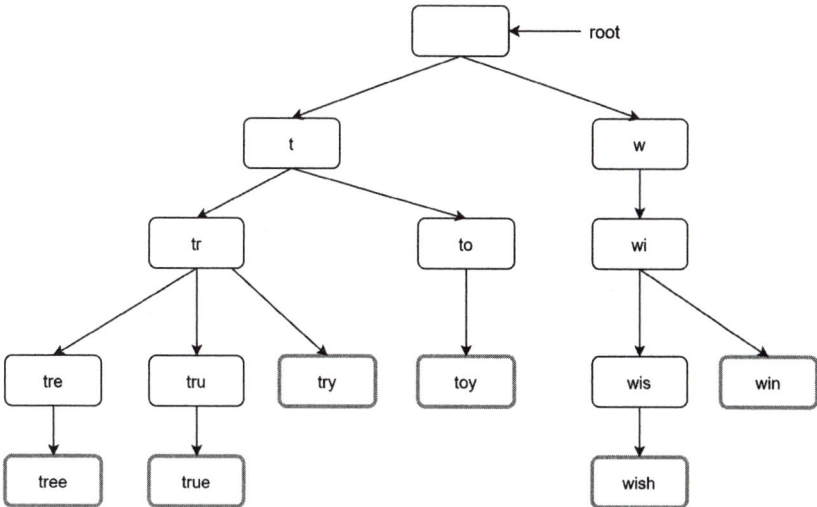

Figure 13-5

Basic trie data structure stores characters in nodes. To support sorting by frequency, frequency info needs to be included in nodes. Assume we have the following frequency table.

Query	Frequency
tree	10
try	29
true	35
toy	14
wish	25
win	50

Table 13-2

After adding frequency info to nodes, updated trie data structure is shown in Figure 13-6.

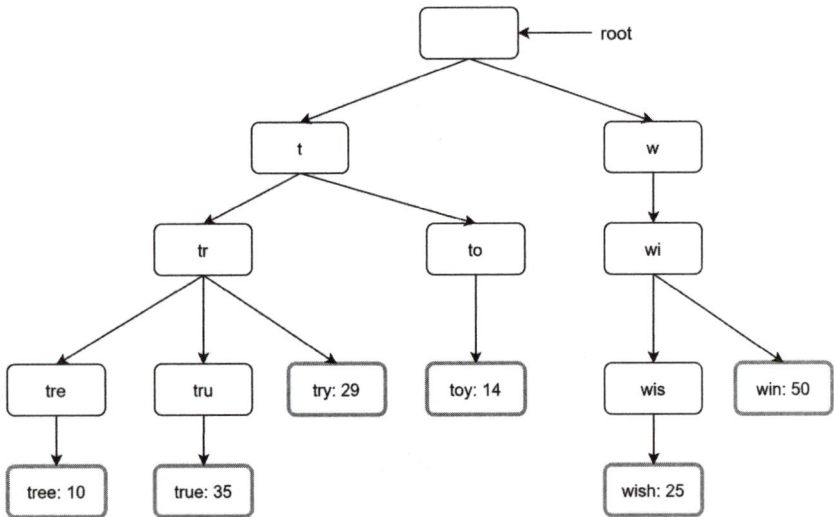

Figure 13-6

How does autocomplete work with trie? Before diving into the algorithm, let us define some terms.

- p: length of a prefix
- n: total number of nodes in a trie
- c: number of children of a given node

Steps to get top *k* most searched queries are listed below:

1. Find the prefix. Time complexity: $O(p)$.

2. Traverse the subtree from the prefix node to get all valid children. A child is valid if it can form a valid query string. Time complexity: $O(c)$

3. Sort the children and get top *k*. Time complexity: $O(c\log c)$

Let us use an example as shown in Figure 13-7 to explain the algorithm. Assume *k* equals to 2 and a user types "tr" in the search box. The algorithm works as follows:

- Step 1: Find the prefix node "tr".

- Step 2: Traverse the subtree to get all valid children. In this case, nodes [tree: 10], [true: 35], [try: 29] are valid.

- Step 3: Sort the children and get top 2. [true: 35] and [try: 29] are the top 2 queries with prefix "tr".

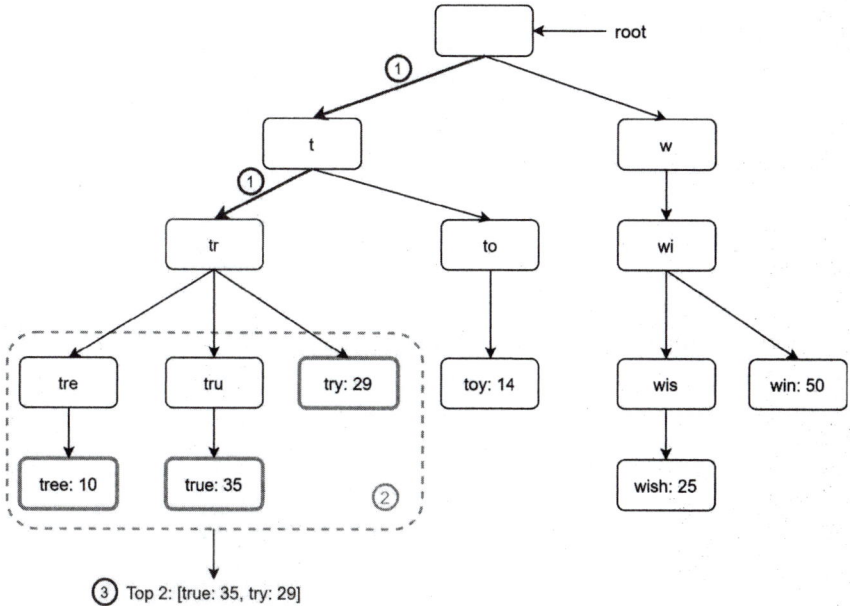

Figure 13-7

The time complexity of this algorithm is the sum of time spent on each step mentioned above: $O(p) + O(c) + O(c\log c)$

The above algorithm is straightforward. However, it is too slow because we need to traverse the entire trie to get top k results in the worst-case scenario. Below are two optimizations:

1. Limit the max length of a prefix
2. Cache top search queries at each node

Let us look at these optimizations one by one.

Limit the max length of a prefix

Users rarely type a long search query into the search box. Thus, it is safe to say p is a small integer number, say 50. If we limit the length of a prefix, the time complexity for "Find the prefix" can be reduced from $O(p)$ to $O(small\ constant)$, aka $O(1)$.

Cache top search queries at each node

To avoid traversing the whole trie, we store top k most frequently used queries at each node. Since 5 to 10 autocomplete suggestions are enough for users, k is a relatively small number. In our specific case, only the top 5 search queries are cached.

By caching top search queries at every node, we significantly reduce the time complexity to retrieve the top 5 queries. However, this design requires a lot of space to store top queries at every node. Trading space for time is well worth it as fast response time is very important.

Figure 13-8 shows the updated trie data structure. Top 5 queries are stored on each node. For example, the node with prefix "be" stores the following: [best: 35, bet: 29, bee: 20, be: 15, beer: 10].

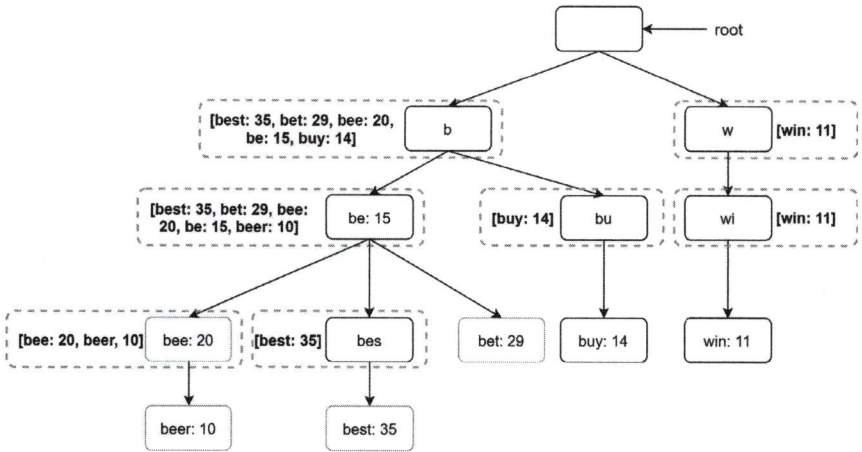

Figure 13-8

Let us revisit the time complexity of the algorithm after applying those two optimizations:

1. Find the prefix node. Time complexity: *O(1)*

2. Return top *k*. Since top *k* queries are cached, the time complexity for this step is *O(1)*.

As the time complexity for each of the steps is reduced to *O(1)*, our algorithm takes only *O(1)* to fetch top *k* queries.

Data gathering service

In our previous design, whenever a user types a search query, data is updated in real-time. This approach is not practical for the following two reasons:

- Users may enter billions of queries per day. Updating the trie on every query significantly slows down the query service.

- Top suggestions may not change much once the trie is built. Thus, it is unnecessary to update the trie frequently.

To design a scalable data gathering service, we examine where data comes from and how data is used. Real-time applications like Twitter require up to date autocomplete suggestions. However, autocomplete suggestions for many Google keywords might not change much on a daily basis.

Despite the differences in use cases, the underlying foundation for data gathering service remains the same because data used to build the trie is usually from analytics or logging services.

Figure 13-9 shows the redesigned data gathering service. Each component is examined one by one.

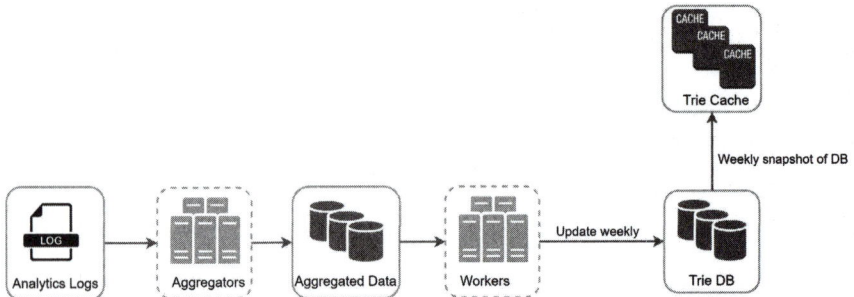

Figure 13-9

Analytics Logs. It stores raw data about search queries. Logs are append-only and are not indexed. Table 13-3 shows an example of the log file.

query	time
tree	2019-10-01 22:01:01
try	2019-10-01 22:01:05
tree	2019-10-01 22:01:30
toy	2019-10-01 22:02:22
tree	2019-10-02 22:02:42
try	2019-10-03 22:03:03

Table 13-3

Aggregators. The size of analytics logs is usually very large, and data is not in the right format. We need to aggregate data so it can be easily processed by our system.

Depending on the use case, we may aggregate data differently. For real-time applications such as Twitter, we aggregate data in a shorter time interval as real-time results are important. On the other hand, aggregating data less frequently, say once per week, might be good enough for many use cases. During an interview session, verify whether real-time results are important. We assume trie is rebuilt weekly.

Aggregated Data.

Table 13-4 shows an example of aggregated weekly data. "time" field represents the start time of a week. "frequency" field is the sum of the occurrences for the corresponding query in that week.

query	time	frequency
tree	2019-10-01	12000
tree	2019-10-08	15000
tree	2019-10-15	9000
toy	2019-10-01	8500
toy	2019-10-08	6256
toy	2019-10-15	8866

Table 13-4

Workers. Workers are a set of servers that perform asynchronous jobs at regular intervals. They build the trie data structure and store it in Trie DB.

Trie Cache. Trie Cache is a distributed cache system that keeps trie in memory for fast read. It takes a weekly snapshot of the DB.

Trie DB. Trie DB is the persistent storage. Two options are available to store the data:

1. Document store: Since a new trie is built weekly, we can periodi-

cally take a snapshot of it, serialize it, and store the serialized data
in the database. Document stores like MongoDB [4] are good
fits for serialized data.

2. Key-value store: A trie can be represented in a hash table form [4]
    by applying the following logic:

    •   Every prefix in the trie is mapped to a key in a hash table.

    •   Data on each trie node is mapped to a value in a hash table.

Figure 13-10 shows the mapping between the trie and hash table.

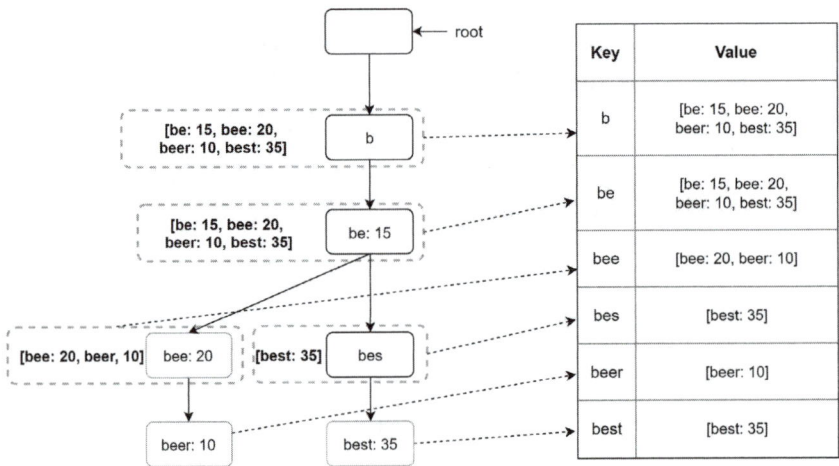

Key	Value
b	[be: 15, bee: 20, beer: 10, best: 35]
be	[be: 15, bee: 20, beer: 10, best: 35]
bee	[bee: 20, beer: 10]
bes	[best: 35]
beer	[beer: 10]
best	[best: 35]

Figure 13-10

In Figure 13-10, each trie node on the left is mapped to the *<key, value>*
pair on the right. If you are unclear how key-value stores work, refer to
Chapter 6: Design a key-value store.

Query service

In the high-level design, query service calls the database directly to fetch
the top 5 results. Figure 13-11 shows the improved design as previous
design is inefficient.

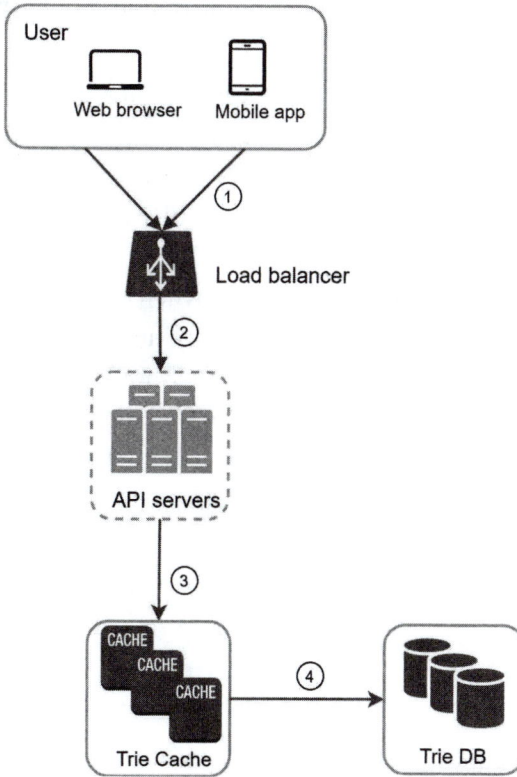

Figure 13-11

1. A search query is sent to the load balancer.

2. The load balancer routes the request to API servers.

3. API servers get trie data from Trie Cache and construct autocomplete suggestions for the client.

4. In case the data is not in Trie Cache, we replenish data back to the cache. This way, all subsequent requests for the same prefix are returned from the cache. A cache miss can happen when a cache server is out of memory or offline.

Query service requires lightning-fast speed. We propose the following optimizations:

- AJAX request. For web applications, browsers usually send AJAX requests to fetch autocomplete results. The main benefit of AJAX is that sending/receiving a request/response does not refresh the whole web page.

- Browser caching. For many applications, autocomplete search suggestions may not change much within a short time. Thus, autocomplete suggestions can be saved in browser cache to allow subsequent requests to get results from the cache directly. Google search engine uses the same cache mechanism. Figure 13-12 shows the response header when you type "system design interview" on the Google search engine. As you can see, Google caches the results in the browser for 1 hour. Please note: "private" in cache-control means results are intended for a single user and must not be cached by a shared cache. "max-age=3600" means the cache is valid for 3600 seconds, aka, an hour.

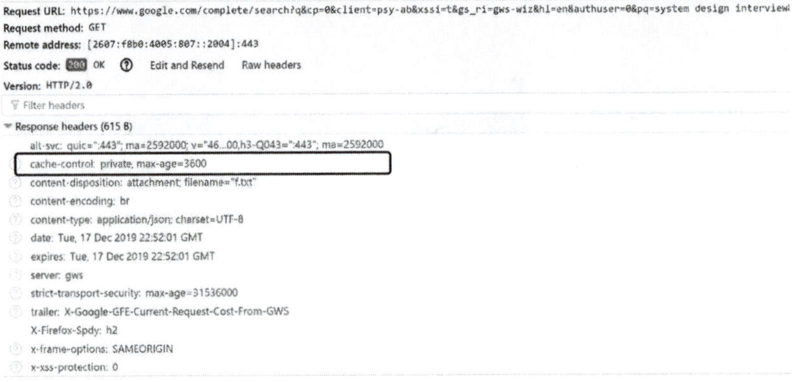

Request URL: https://www.google.com/complete/search?q&cp=0&client=psy-ab&xssi=t&gs_ri=gws-wiz&hl=en&authuser=0&pq=system design interview
Request method: GET
Remote address: [2607:f8b0:4005:807::2004]:443
Status code: 200 OK ⓘ   Edit and Resend   Raw headers
Version: HTTP/2.0
⛛ Filter headers
▾ Response headers (615 B)
    alt-svc: quic=":443"; ma=2592000; v="46...00,h3-Q043=":443"; ma=2592000
    cache-control: private, max-age=3600
    content-disposition: attachment; filename="f.txt"
    content-encoding: br
    content-type: application/json; charset=UTF-8
    date: Tue, 17 Dec 2019 22:52:01 GMT
    expires: Tue, 17 Dec 2019 22:52:01 GMT
    server: gws
    strict-transport-security: max-age=31536000
    trailer: X-Google-GFE-Current-Request-Cost-From-GWS
    X-Firefox-Spdy: h2
    x-frame-options: SAMEORIGIN
    x-xss-protection: 0

Figure 13-12

- Data sampling: For a large-scale system, logging every search query requires a lot of processing power and storage. Data sampling is important. For instance, only 1 out of every N requests is logged by the system.

Trie operations

Trie is a core component of the autocomplete system. Let us look at how trie operations (create, update, and delete) work.

Create

Trie is created by workers using aggregated data. The source of data is from Analytics Log/DB.

Update

There are two ways to update the trie.

Option 1: Update the trie weekly. Once a new trie is created, the new trie replaces the old one.

Option 2: Update individual trie node directly. We try to avoid this operation because it is slow. However, if the size of the trie is small, it is an acceptable solution. When we update a trie node, its ancestors all the way up to the root must be updated because ancestors store top queries of children. Figure 13-13 shows an example of how the update operation works. On the left side, the search query "beer" has the original value 10. On the right side, it is updated to 30. As you can see, the node and its ancestors have the "beer" value updated to 30.

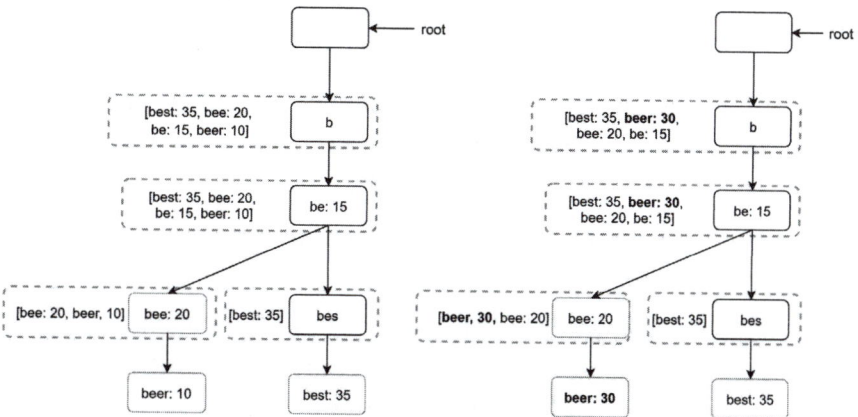

Figure 13-13

Delete

We have to remove hateful, violent, sexually explicit, or dangerous autocomplete suggestions. We add a filter layer (Figure 13-14) in front of the Trie Cache to filter out unwanted suggestions. Having a filter layer gives us the flexibility of removing results based on different filter rules. Unwanted suggestions are removed physically from the database asynchronically so the correct data set will be used to build trie in the next update cycle.

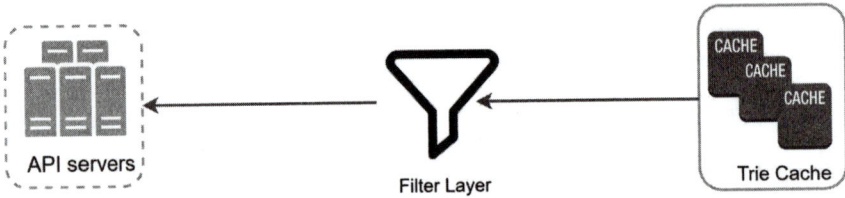

Figure 13-14

Scale the storage

Now that we have developed a system to bring autocomplete queries to users, it is time to solve the scalability issue when the trie grows too large to fit in one server.

Since English is the only supported language, a naive way to shard is based on the first character. Here are some examples.

- If we need two servers for storage, we can store queries starting with 'a' to 'm' on the first server, and 'n' to 'z' on the second server.

- If we need three servers, we can split queries into 'a' to 'i', 'j' to 'r' and 's' to 'z'.

Following this logic, we can split queries up to 26 servers because there are 26 alphabetic characters in English. Let us define sharding based on the first character as first level sharding. To store data beyond 26 servers, we can shard on the second or even at the third level. For example, data queries that start with 'a' can be split into 4 servers: 'aa-ag', 'ah-an', 'ao-au', and 'av-az'.

At the first glance this approach seems reasonable, until you realize that there are a lot more words that start with the letter 'c' than 'x'. This creates uneven distribution.

To mitigate the data imbalance problem, we analyze historical data distribution pattern and apply smarter sharding logic as shown in Figure 13-15. The shard map manager maintains a lookup database for identifying where rows should be stored. For example, if there are a similar number of historical queries for 's' and for 'u', 'v', 'w', 'x', 'y' and 'z' combined, we can maintain two shards: one for 's' and one for 'u' to 'z'.

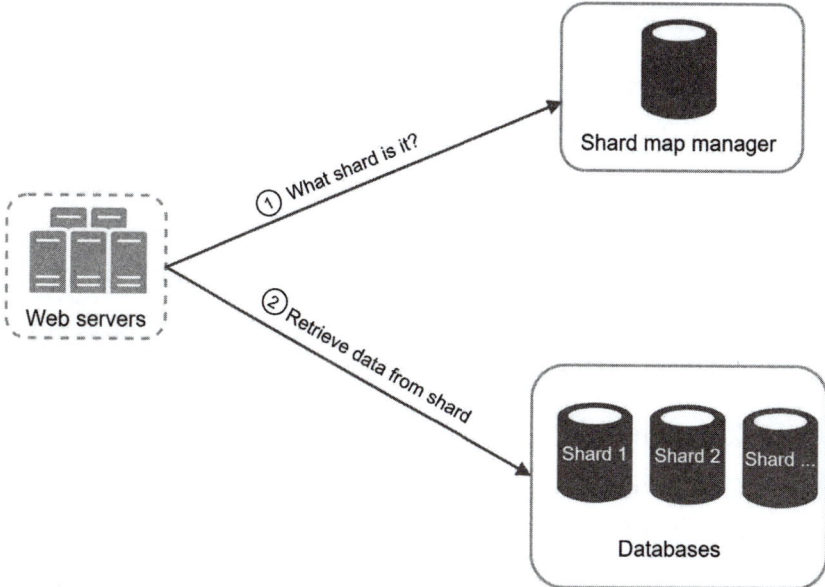

Figure 13-15

Step 4 - Wrap up

After you finish the deep dive, your interviewer might ask you some follow up questions.

Interviewer: How do you extend your design to support multiple languages?

To support other non-English queries, we store Unicode characters in trie nodes. If you are not familiar with Unicode, here is the definition: "an encoding standard covers all the characters for all the writing systems of the world, modern and ancient" [5].

Interviewer: What if top search queries in one country are different from others?

In this case, we might build different tries for different countries. To improve the response time, we can store tries in CDNs.

Interviewer: How can we support the trending (real-time) search queries?

Assuming a news event breaks out, a search query suddenly becomes popular. Our original design will not work because:

- Offline workers are not scheduled to update the trie yet because this is scheduled to run on weekly basis.
- Even if it is scheduled, it takes too long to build the trie.

Building a real-time search autocomplete is complicated and is beyond the scope of this book so we will only give a few ideas:

- Reduce the working data set by sharding.
- Change the ranking model and assign more weight to recent search queries.
- Data may come as streams, so we do not have access to all the data at once. Streaming data means data is generated continuously. Stream processing requires a different set of systems: Apache Hadoop MapReduce [6], Apache Spark Streaming [7], Apache Storm [8], Apache Kafka [9], etc. Because all those topics require specific domain knowledge, we are not going into detail here.

Congratulations on getting this far! Now give yourself a pat on the back. Good job!

Reference materials

[1] The Life of a Typeahead Query:
https://www.facebook.com/notes/facebook-engineering/the-life-of-a-
typeahead-query/389105248919/

[2] How We Built Prefixy: A Scalable Prefix Search Service for Powering
Autocomplete:
https://medium.com/@prefixyteam/how-we-built-prefixy-a-scalable-
prefix-search-service-for-powering-autocomplete-c20f98e2eff1

[3] Prefix Hash Tree An Indexing Data Structure over Distributed
Hash Tables:
https://people.eecs.berkeley.edu/~sylvia/papers/pht.pdf

[4] MongoDB wikipedia: https://en.wikipedia.org/wiki/MongoDB

[5] Unicode frequently asked questions:
https://www.unicode.org/faq/basic_q.html

[6] Apache hadoop: https://hadoop.apache.org/

[7] Spark streaming: https://spark.apache.org/streaming/

[8] Apache storm: https://storm.apache.org/

[9] Apache kafka: https://kafka.apache.org/documentation/

14

DESIGN YOUTUBE

In this chapter, you are asked to design YouTube. The solution to this question can be applied to other interview questions like designing a video sharing platform such as Netflix and Hulu. Figure 14-1 shows the YouTube homepage.

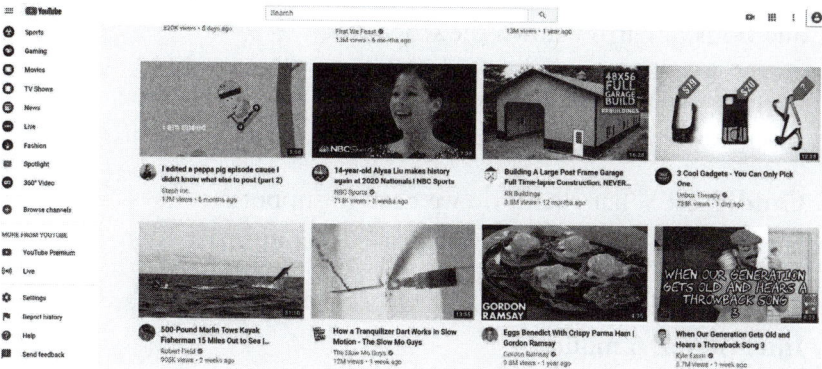

Figure 14-1

YouTube looks simple: content creators upload videos and viewers click play. Is it really that simple? Not really. There are lots of complex technologies underneath the simplicity. Let us look at some impressive statistics, demographics, and fun facts of YouTube in 2020 [1] [2].

- Total number of monthly active users: 2 billion.
- Number of videos watched per day: 5 billion.
- 73% of US adults use YouTube.
- 50 million creators on YouTube
- YouTube's Ad revenue was $15.1 billion for the full year 2019, up 36% from 2018.

- YouTube is responsible for 37% of all mobile internet traffic.
- YouTube is available in 80 different languages.

From these statistics, we know YouTube is enormous, global and makes a lot of money.

Step I - Understand the problem and establish design scope

As revealed in Figure 14-1, besides watching a video, you can do a lot more on YouTube. For example, comment, share, or like a video, save a video to playlists, subscribe to a channel, etc. It is impossible to design everything within a 45- or 60-minute interview. Thus, it is important to ask questions to narrow down the scope.

Candidate: What features are important?
Interviewer: Ability to upload a video and watch a video.

Candidate: What clients do we need to support?
Interviewer: Mobile apps, web browsers, and smart TV.

Candidate: How many daily active users do we have?
Interviewer: 5 million

Candidate: What is the average daily time spent on the product?
Interviewer: 30 minutes.

Candidate: Do we need to support international users?
Interviewer: Yes, a large percentage of users are international users.

Candidate: What are the supported video resolutions?
Interviewer: The system accepts most of the video resolutions and formats.

Candidate: Is encryption required?
Interviewer: Yes

Candidate: Any file size requirement for videos?
Interviewer: Our platform focuses on small and medium-sized videos. The maximum allowed video size is 1GB.

Candidate: Can we leverage some of the existing cloud infrastructures provided by Amazon, Google, or Microsoft?
Interviewer: That is a great question. Building everything from scratch is unrealistic for most companies, it is recommended to leverage some of the existing cloud services.

In the chapter, we focus on designing a video streaming service with the following features:

- Ability to upload videos fast

- Smooth video streaming

- Ability to change video quality

- Low infrastructure cost

- High availability, scalability, and reliability requirements

- Clients supported: mobile apps, web browser, and smart TV

Back of the envelope estimation

The following estimations are based on many assumptions, so it is important to communicate with the interviewer to make sure she is on the same page.

- Assume the product has 5 million daily active users (DAU).

- Users watch 5 videos per day.

- 10% of users upload 1 video per day.

- Assume the average video size is 300 MB.

- Total daily storage space needed: 5 million * 10% * 300 MB = 150TB

- CDN cost.

  o   When cloud CDN serves a video, you are charged for data transferred out of the CDN.

  o   Let us use Amazon's CDN CloudFront for cost estimation (Figure 14-2) [3]. Assume 100% of traffic is served from the

United States. The average cost per GB is $0.02. For simplicity, we only calculate the cost of video streaming.

o   5 million * 5 videos * 0.3GB * $0.02 = $150,000 per day.

From the rough cost estimation, we know serving videos from the CDN costs lots of money. Even though cloud providers are willing to lower the CDN costs significantly for big customers, the cost is still substantial. We will discuss ways to reduce CDN costs in deep dive.

Per Month	United States & Canada	Europe & Israel	South Africa, Kenya, & Middle East	South America	Japan	Australia	Singapore, South Korea, Taiwan, Hong Kong, & Philippines	India
First 10TB	$0.085	$0.085	$0.110	$0.110	$0.114	$0.114	$0.140	$0.170
Next 40TB	$0.080	$0.080	$0.105	$0.105	$0.089	$0.098	$0.135	$0.130
Next 100TB	$0.060	$0.060	$0.090	$0.090	$0.086	$0.094	$0.120	$0.110
Next 350TB	$0.040	$0.040	$0.080	$0.080	$0.084	$0.092	$0.100	$0.100
Next 524TB	$0.030	$0.030	$0.060	$0.060	$0.080	$0.090	$0.080	$0.100
Next 4PB	$0.025	$0.025	$0.050	$0.050	$0.070	$0.085	$0.070	$0.100
Over 5PB	$0.020	$0.020	$0.040	$0.040	$0.060	$0.080	$0.060	$0.100

Figure 14-2

Step 2 - Propose high-level design and get buy-in

As discussed previously, the interviewer recommended leveraging existing cloud services instead of building everything from scratch. CDN and blob storage are the cloud services we will leverage. Some readers might ask why not building everything by ourselves? Reasons are listed below:

- System design interviews are not about building everything from scratch. Within the limited time frame, choosing the right technology to do a job right is more important than explaining how the technology works in detail. For instance, mentioning blob storage for storing source videos is enough for the interview. Talking about the detailed design for blob storage could be an overkill.

- Building scalable blob storage or CDN is extremely complex and costly. Even large companies like Netflix or Facebook do not build everything themselves. Netflix leverages Amazon's cloud services [4], and Facebook uses Akamai's CDN [5].

At the high-level, the system comprises three components (Figure 14-3).

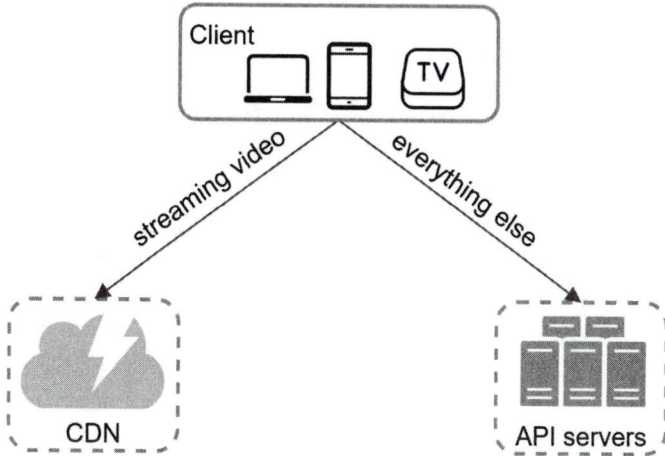

Figure 14-3

Client: You can watch YouTube on your computer, mobile phone, and smartTV.

CDN: Videos are stored in CDN. When you press play, a video is streamed from the CDN.

API servers: Everything else except video streaming goes through API servers. This includes feed recommendation, generating video upload URL, updating metadata database and cache, user signup, etc.

In the question/answer session, the interviewer showed interests in two flows:

- Video uploading flow
- Video streaming flow

We will explore the high-level design for each of them.

Video uploading flow

Figure 14-4 shows the high-level design for the video uploading.

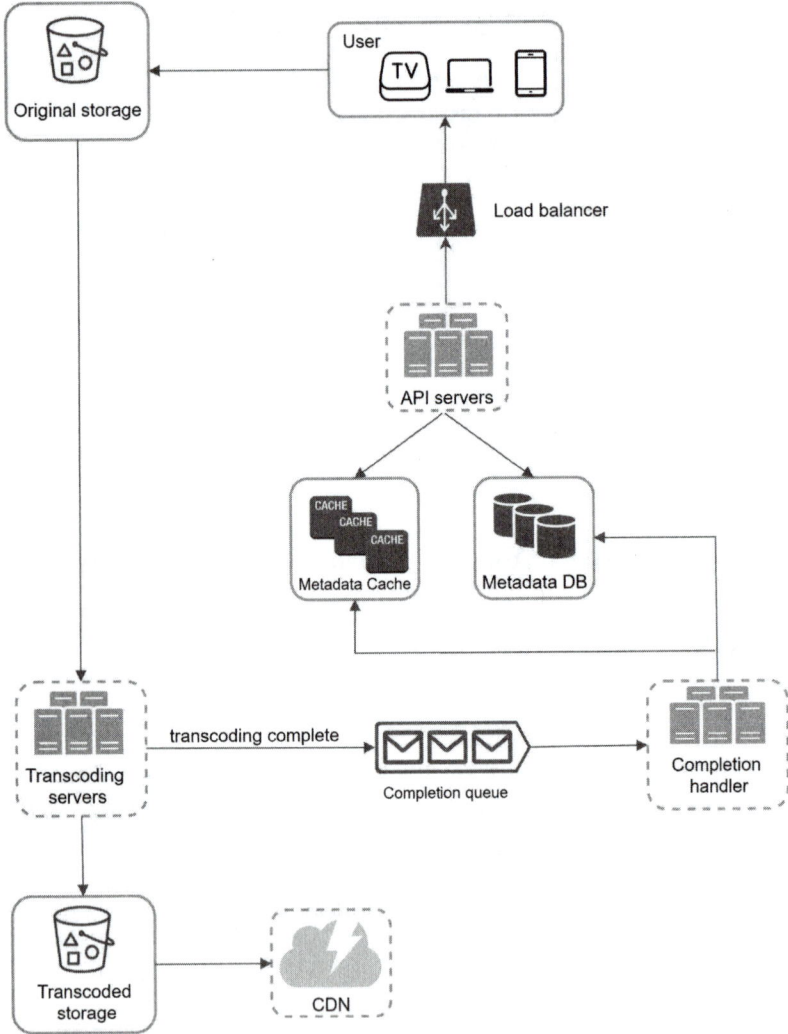

Figure 14-4

It consists of the following components:

- User: A user watches YouTube on devices such as a computer, mobile phone, or smart TV.

- Load balancer: A load balancer evenly distributes requests among API servers.

- API servers: All user requests go through API servers except video streaming.

- Metadata DB: Video metadata are stored in Metadata DB. It is sharded and replicated to meet performance and high availability requirements.

- Metadata cache: For better performance, video metadata and user objects are cached.

- Original storage: A blob storage system is used to store original videos. A quotation in Wikipedia regarding blob storage shows that: "A Binary Large Object (BLOB) is a collection of binary data stored as a single entity in a database management system" [6].

- Transcoding servers: Video transcoding is also called video encoding. It is the process of converting a video format to other formats (MPEG, HLS, etc), which provide the best video streams possible for different devices and bandwidth capabilities.

- Transcoded storage: It is a blob storage that stores transcoded video files.

- CDN: Videos are cached in CDN. When you click the play button, a video is streamed from the CDN.

- Completion queue: It is a message queue that stores information about video transcoding completion events.

- Completion handler: This consists of a list of workers that pull event data from the completion queue and update metadata cache and database.

Now that we understand each component individually, let us examine how the video uploading flow works. The flow is broken down into two processes running in parallel.

a. Upload the actual video.

b. Update video metadata. Metadata contains information about video URL, size, resolution, format, user info, etc.

Flow a: upload the actual video

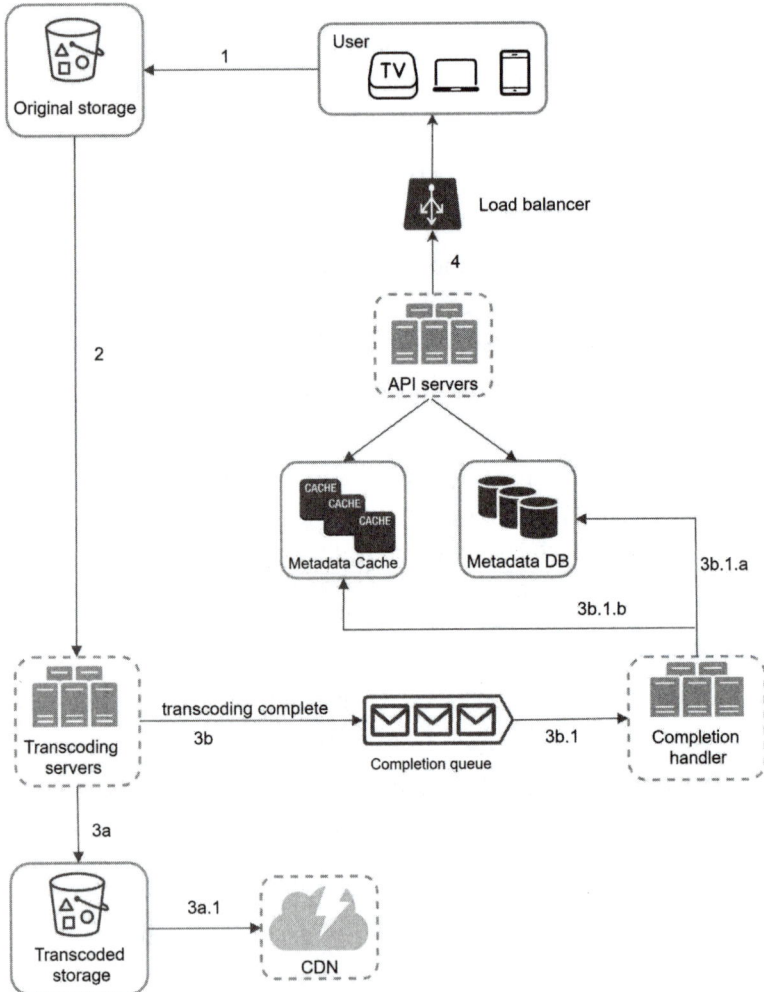

Figure 14-5

Figure 14-5 shows how to upload the actual video. The explanation is shown below:

1. Videos are uploaded to the original storage.

2. Transcoding servers fetch videos from the original storage and start transcoding.

3. Once transcoding is complete, the following two steps are executed in parallel:

   3a. Transcoded videos are sent to transcoded storage.

   3b. Transcoding completion events are queued in the completion queue.

3a.1. Transcoded videos are distributed to CDN.

3b.1. Completion handler contains a bunch of workers that continuously pull event data from the queue.

3b.1.a. and 3b.1.b. Completion handler updates the metadata database and cache when video transcoding is complete.

4. API servers inform the client that the video is successfully uploaded and is ready for streaming.

Flow b: update the metadata

While a file is being uploaded to the original storage, the client in parallel sends a request to update the video metadata as shown in Figure 14-6. The request contains video metadata, including file name, size, format, etc. API servers update the metadata cache and database.

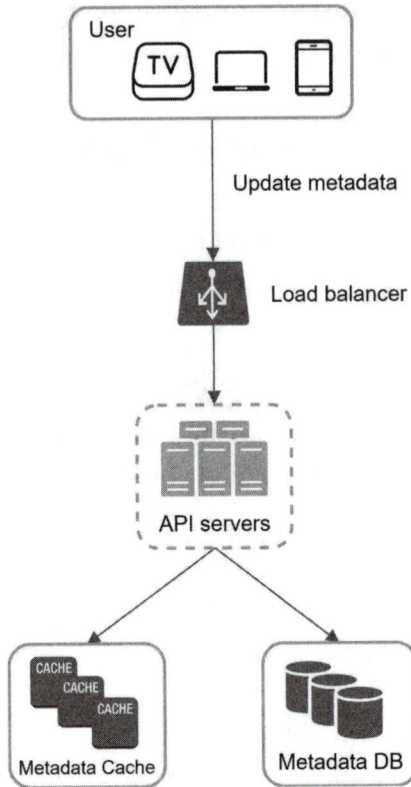

Figure 14-6

Video streaming flow

Whenever you watch a video on YouTube, it usually starts streaming immediately and you do not wait until the whole video is downloaded. Downloading means the whole video is copied to your device, while streaming means your device continuously receives video streams from remote source videos. When you watch streaming videos, your client loads a little bit of data at a time so you can watch videos immediately and continuously.

Before we discuss video streaming flow, let us look at an important concept: streaming protocol. This is a standardized way to control data transfer for video streaming. Popular streaming protocols are:

- MPEG–DASH. MPEG stands for "Moving Picture Experts Group" and DASH stands for "Dynamic Adaptive Streaming over HTTP".

- Apple HLS. HLS stands for "HTTP Live Streaming".

- Microsoft Smooth Streaming.

- Adobe HTTP Dynamic Streaming (HDS).

You do not need to fully understand or even remember those streaming protocol names as they are low-level details that require specific domain knowledge. The important thing here is to understand that different streaming protocols support different video encodings and playback players. When we design a video streaming service, we have to choose the right streaming protocol to support our use cases. To learn more about streaming protocols, here is an excellent article [7].

Videos are streamed from CDN directly. The edge server closest to you will deliver the video. Thus, there is very little latency. Figure 14-7 shows a high level of design for video streaming.

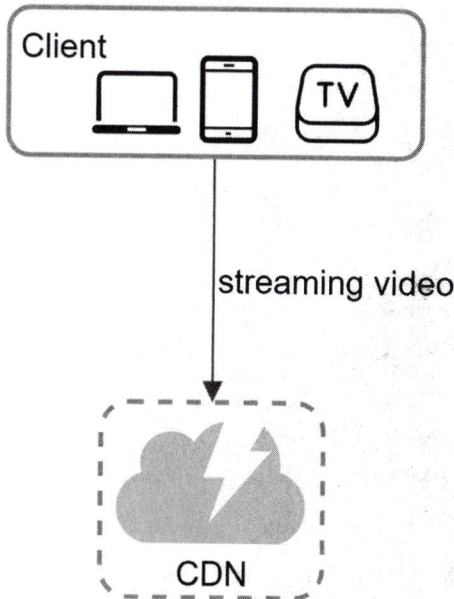

Figure 14-7

Step 3 - Design deep dive

In the high-level design, the entire system is broken down in two parts: video uploading flow and video streaming flow. In this section, we will refine both flows with important optimizations and introduce error handling mechanisms.

Video transcoding

When you record a video, the device (usually a phone or camera) gives the video file a certain format. If you want the video to be played smoothly on other devices, the video must be encoded into compatible bitrates and formats. Bitrate is the rate at which bits are processed over time. A higher bitrate generally means higher video quality. High bitrate streams need more processing power and fast internet speed.

Video transcoding is important for the following reasons:

- Raw video consumes large amounts of storage space. An hour-long high definition video recorded at 60 frames per second can take up a few hundred GB of space.

- Many devices and browsers only support certain types of video formats. Thus, it is important to encode a video to different formats for compatibility reasons.

- To ensure users watch high-quality videos while maintaining smooth playback, it is a good idea to deliver higher resolution video to users who have high network bandwidth and lower resolution video to users who have low bandwidth.

- Network conditions can change, especially on mobile devices. To ensure a video is played continuously, switching video quality automatically or manually based on network conditions is essential for smooth user experience.

Many types of encoding formats are available; however, most of them contain two parts:

- Container: This is like a basket that contains the video file, audio, and metadata. You can tell the container format by the file extension, such as .avi, .mov, or .mp4.

- Codecs: These are compression and decompression algorithms aim to reduce the video size while preserving the video quality. The most used video codecs are H.264, VP9, and HEVC.

Directed acyclic graph (DAG) model

Transcoding a video is computationally expensive and time-consuming. Besides, different content creators may have different video processing requirements. For instance, some content creators require watermarks on top of their videos, some provide thumbnail images themselves, and some upload high definition videos, whereas others do not.

To support different video processing pipelines and maintain high parallelism, it is important to add some level of abstraction and let client programmers define what tasks to execute. For example, Facebook's streaming video engine uses a directed acyclic graph (DAG) programming model, which defines tasks in stages so they can be executed sequentially or parallelly [8]. In our design, we adopt a similar DAG model to achieve flexibility and parallelism. Figure 14-8 represents a DAG for video transcoding.

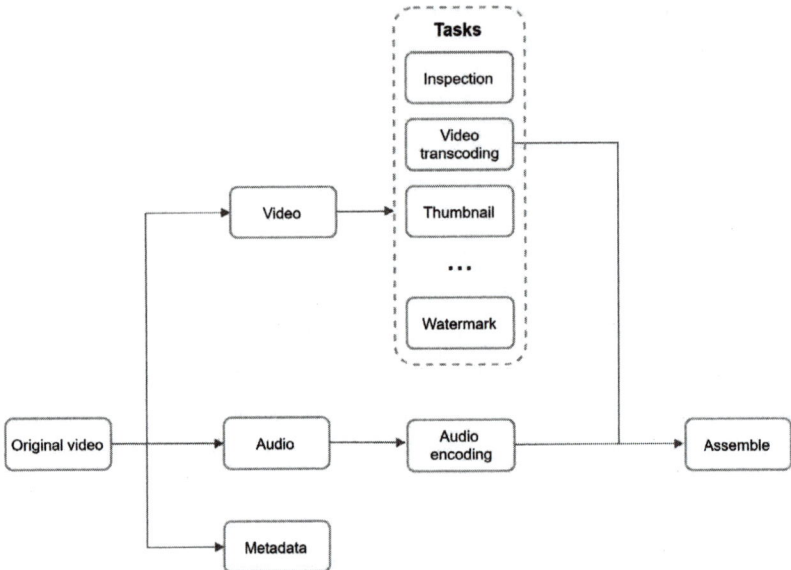

Figure 14-8

In Figure 14-8, the original video is split into video, audio, and metadata. Here are some of the tasks that can be applied on a video file:

- Inspection: Make sure videos have good quality and are not malformed.

- Video encodings: Videos are converted to support different resolutions, codec, bitrates, etc. Figure 14-9 shows an example of video encoded files.

- Thumbnail. Thumbnails can either be uploaded by a user or automatically generated by the system.

- Watermark: An image overlay on top of your video contains identifying information about your video.

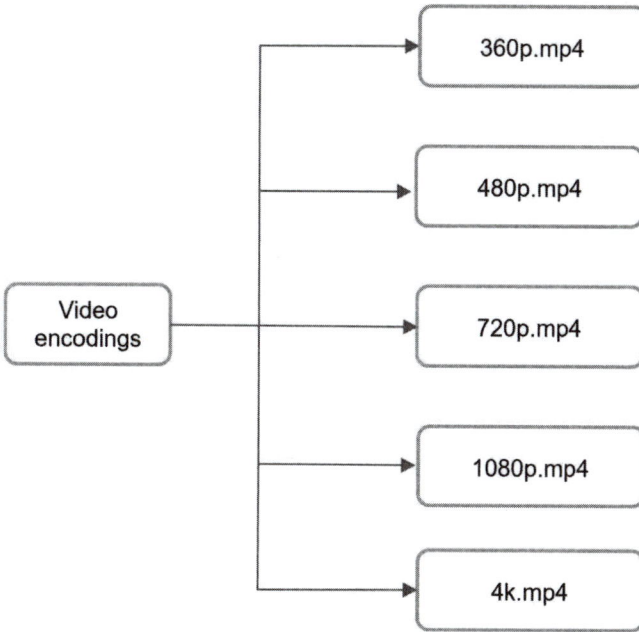

Figure 14-9

Video transcoding architecture

The proposed video transcoding architecture that leverages the cloud services, is shown in Figure 14-10.

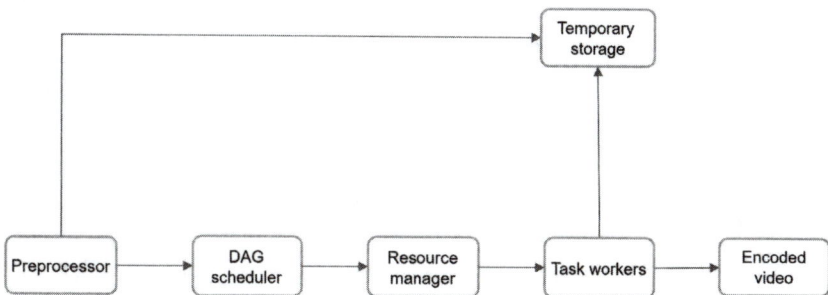

Figure 14-10

The architecture has six main components: preprocessor, DAG scheduler, resource manager, task workers, temporary storage, and encoded video as the output. Let us take a close look at each component.

Preprocessor

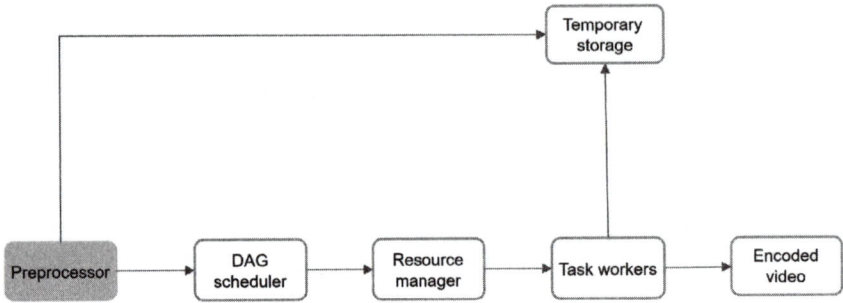

Fiugre 14-11

The preprocessor has 4 responsibilities:

1. Video splitting. Video stream is split or further split into smaller Group of Pictures (GOP) alignment. GOP is a group/chunk of frames arranged in a specific order. Each chunk is an independently playable unit, usually a few seconds in length.

2. Some old mobile devices or browsers might not support video splitting. Preprocessor split videos by GOP alignment for old clients.

3. DAG generation. The processor generates DAG based on configuration files client programmers write. Figure 14-12 is a simplified DAG representation which has 2 nodes and 1 edge:

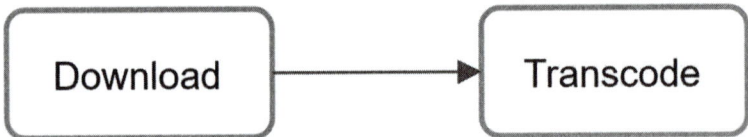

Figure 14-12

This DAG representation is generated from the two configuration files below (Figure 14-13):

```
task {
    name 'download-input'
    type 'Download'
    input {
        url config.url
    }
    output { it->
        context.inputVideo = it.file
    }
    next 'transcode'
}
```

```
task {
    name 'transcode'
    type 'Transcode'
    input {
        input context.inputVideo
        config config.transConfig
    }
    output { it->
        context.file = it.outputVideo
    }
}
```

Figure 14-13 (source: [9])

4. Cache data. The preprocessor is a cache for segmented videos. For better reliability, the preprocessor stores GOPs and metadata in temporary storage. If video encoding fails, the system could use persisted data for retry operations.

DAG scheduler

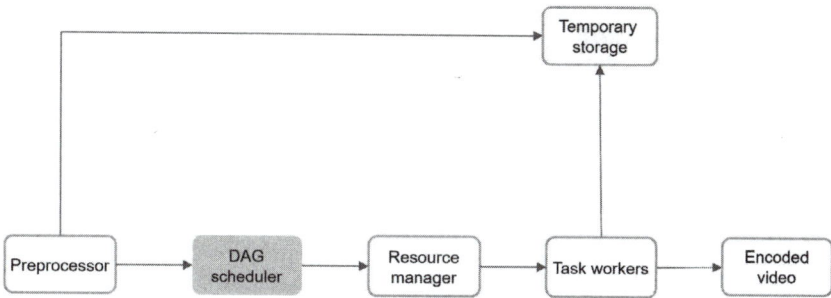

Figure 14-14

The DAG scheduler splits a DAG graph into stages of tasks and puts them in the task queue in the resource manager. Figure 14-15 shows an example of how the DAG scheduler works.

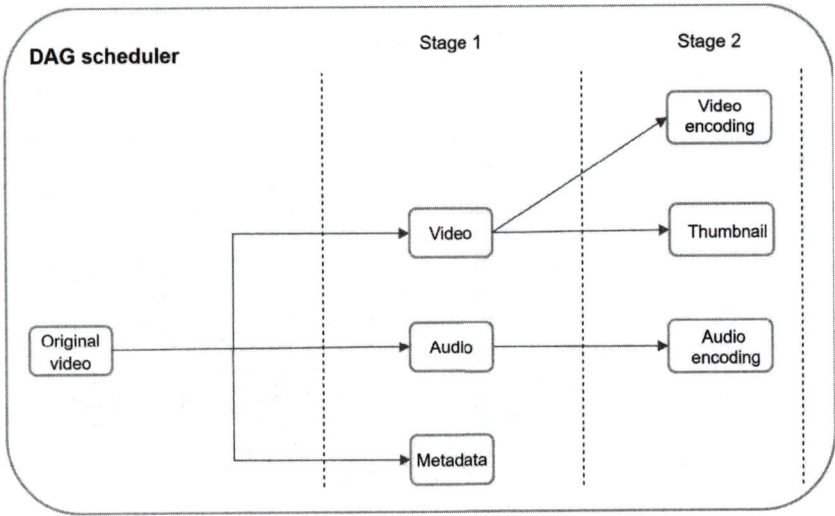

Figure 14-15

As shown in Figure 14-15, the original video is split into 2 stages: Stage 1: video, audio, and metadata. The video file is further split into two tasks in stage 2: video encoding and thumbnail. The audio file requires audio encoding as part of the stage 2 tasks.

Resource manager

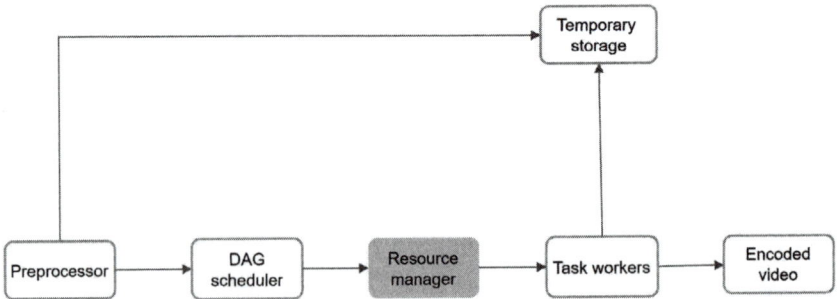

Figure 14-16

The resource manager is responsible for managing the efficiency of resource allocation. It contains 3 queues and a task scheduler as shown in Figure 14-17.

- Task queue: It is a priority queue that contains tasks to be executed.

- Worker queue: It is a priority queue that contains worker utilization info.

- Running queue: It contains info about the currently running tasks and workers running the tasks.

- Task scheduler: It picks the optimal task/worker, and instructs the chosen task worker to execute the job.

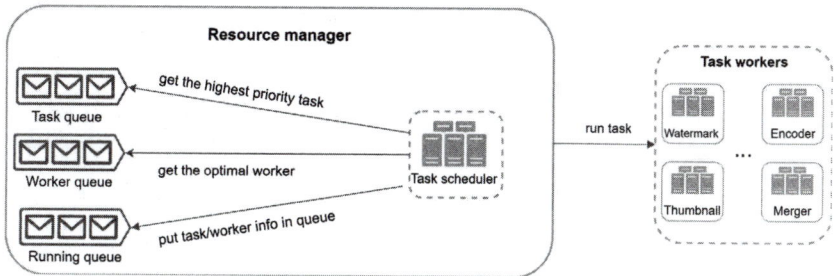

Figure 14-17

The resource manager works as follows:

- The task scheduler gets the highest priority task from the task queue.

- The task scheduler gets the optimal task worker to run the task from the worker queue.

- The task scheduler instructs the chosen task worker to run the task.

- The task scheduler binds the task/worker info and puts it in the running queue.

- The task scheduler removes the job from the running queue once the job is done.

Task workers

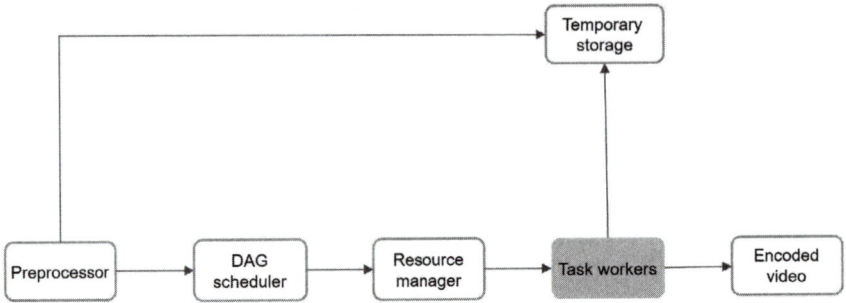

Figure 14-18

Task workers run the tasks which are defined in the DAG. Different task workers may run different tasks as shown in Figure 14-19.

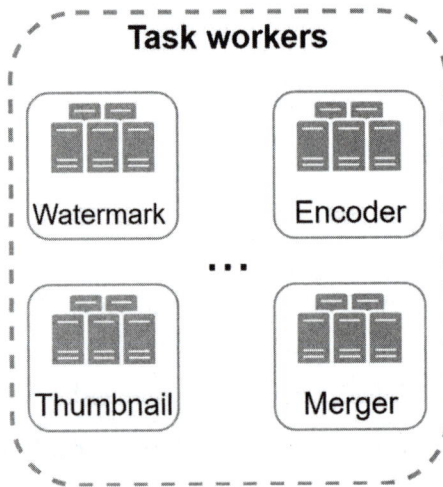

Figure 14-19

Temporary storage

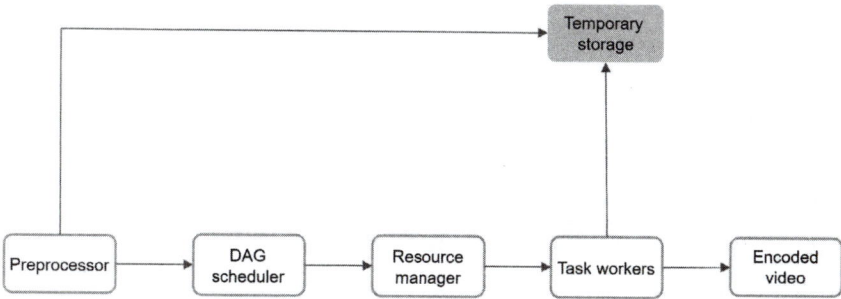

Figure 14-20

Multiple storage systems are used here. The choice of storage system depends on factors like data type, data size, access frequency, data life span, etc. For instance, metadata is frequently accessed by workers, and the data size is usually small. Thus, caching metadata in memory is a good idea. For video or audio data, we put them in blob storage. Data in temporary storage is freed up once the corresponding video processing is complete.

Encoded video

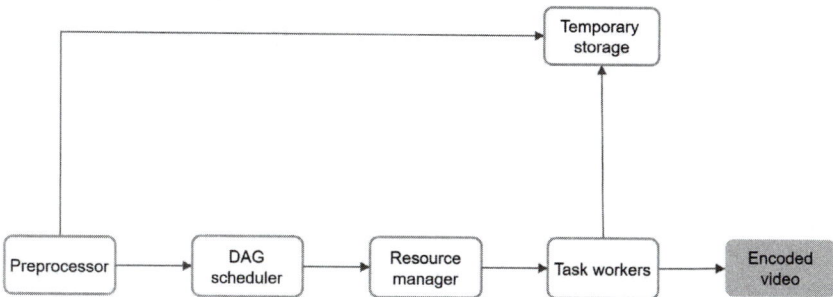

Figure 14-21

Encoded video is the final output of the encoding pipeline. Here is an example of the output: funny_720p.mp4.

System optimizations

At this point, you ought to have good understanding about the video uploading flow, video streaming flow and video transcoding. Next, we will refine the system with optimizations, including speed, safety, and cost-saving.

Speed optimization: parallelize video uploading

Uploading a video as a whole unit is inefficient. We can split a video into smaller chunks by GOP alignment as shown in Figure 14-22.

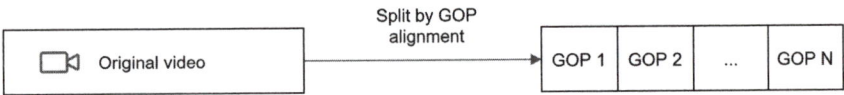

Figure 14-22

This allows fast resumable uploads when the previous upload failed. The job of splitting a video file by GOP can be implemented by the client to improve the upload speed as shown in Figure 14-23.

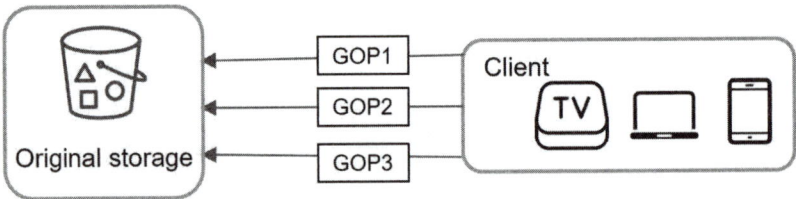

Figure 14-23

Speed optimization: place upload centers close to users

Another way to improve the upload speed is by setting up multiple up-load centers across the globe (Figure 14-24). People in the United States can upload videos to the North America upload center, and people in China can upload videos to the Asian upload center. To achieve this, we use CDN as upload centers.

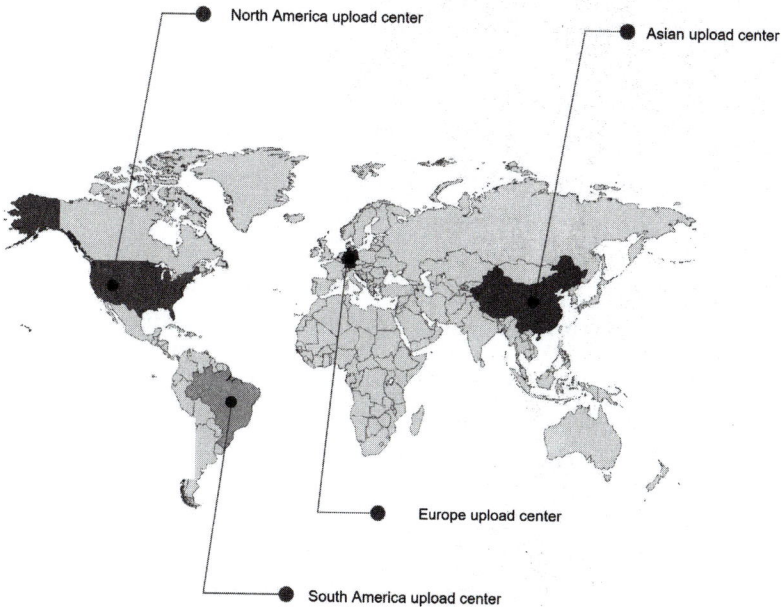

Figure 14-24

Speed optimization: parallelism everywhere

Achieving low latency requires serious efforts. Another optimization is to build a loosely coupled system and enable high parallelism.

Our design needs some modifications to achieve high parallelism. Let us zoom in to the flow of how a video is transferred from original storage to the CDN. The flow is shown in Figure 14-25, revealing that the output depends on the input of the previous step. This dependency makes par-allelism difficult.

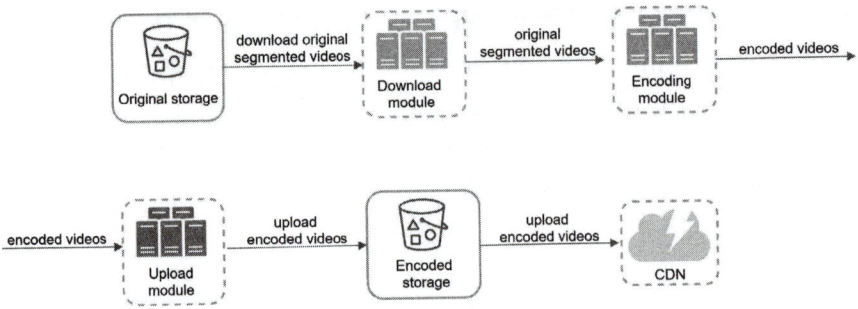

Figure 14-25

To make the system more loosely coupled, we introduced message queues as shown in Figure 14-26. Let us use an example to explain how message queues make the system more loosely coupled.

- Before the message queue is introduced, the encoding module must wait for the output of the download module.

- After the message queue is introduced, the encoding module does not need to wait for the output of the download module anymore. If there are events in the message queue, the encoding module can execute those jobs in parallel.

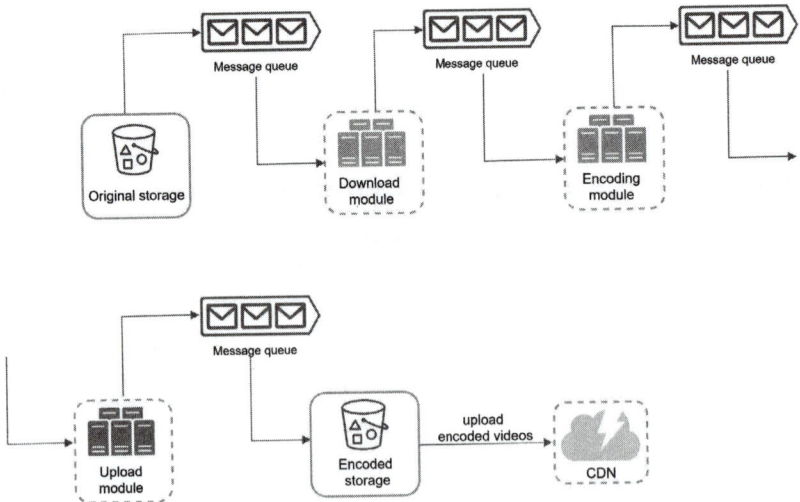

Figure 14-26

Safety optimization: pre-signed upload URL

Safety is one of the most important aspects of any product. To ensure only authorized users upload videos to the right location, we introduce pre-signed URLs as shown in Figure 14-27.

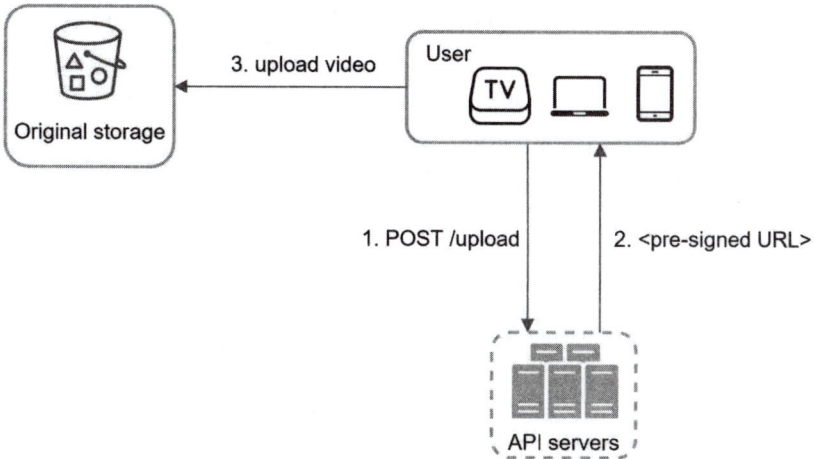

Figure 14-27

The upload flow is updated as follows:

1. The client makes a HTTP request to API servers to fetch the pre-signed URL, which gives the access permission to the object identified in the URL. The term pre-signed URL is used by uploading files to Amazon S3. Other cloud service providers might use a different name. For instance, Microsoft Azure blob storage supports the same feature, but call it "Shared Access Signature" [10].

2. API servers respond with a pre-signed URL.

3. Once the client receives the response, it uploads the video using the pre-signed URL.

Safety optimization: protect your videos

Many content makers are reluctant to post videos online because they fear their original videos will be stolen. To protect copyrighted videos, we can adopt one of the following three safety options:

- Digital rights management (DRM) systems: Three major DRM systems are Apple FairPlay, Google Widevine, and Microsoft PlayReady.

- AES encryption: You can encrypt a video and configure an authorization policy. The encrypted video will be decrypted upon playback. This ensures that only authorized users can watch an encrypted video.

- Visual watermarking: This is an image overlay on top of your video that contains identifying information for your video. It can be your company logo or company name.

Cost-saving optimization

CDN is a crucial component of our system. It ensures fast video delivery on a global scale. However, from the back of the envelope calculation, we know CDN is expensive, especially when the data size is large. How can we reduce the cost?

Previous research shows that YouTube video streams follow long-tail distribution [11] [12]. It means a few popular videos are accessed frequently but many others have few or no viewers. Based on this observation, we implement a few optimizations:

1. Only serve the most popular videos from CDN and other videos from our high capacity storage video servers (Figure 14-28).

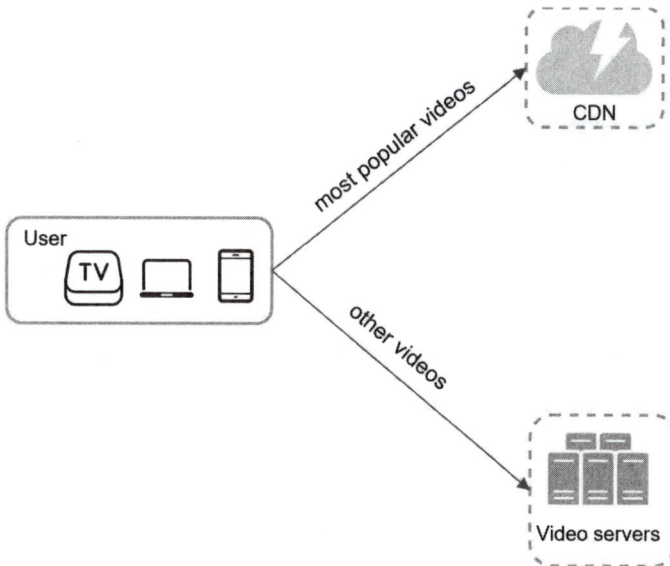

Figure 14-28

2. For less popular content, we may not need to store many encoded video versions. Short videos can be encoded on-demand.

3. Some videos are popular only in certain regions. There is no need to distribute these videos to other regions.

4. Build your own CDN like Netflix and partner with Internet Service Providers (ISPs). Building your CDN is a giant project; however, this could make sense for large streaming companies. An ISP can be Comcast, AT&T, Verizon, or other internet providers. ISPs are located all around the world and are close to users. By partnering with ISPs, you can improve the viewing experience and reduce the bandwidth charges.

All those optimizations are based on content popularity, user access pattern, video size, etc. It is important to analyze historical viewing patterns before doing any optimization. Here are some of the interesting articles on this topic: [12] [13].

Error handling

For a large-scale system, system errors are unavoidable. To build a highly fault-tolerant system, we must handle errors gracefully and recover from them fast. Two types of errors exist:

- Recoverable error. For recoverable errors such as video segment fails to transcode, the general idea is to retry the operation a few times. If the task continues to fail and the system believes it is not recoverable, it returns a proper error code to the client.

- Non-recoverable error. For non-recoverable errors such as malformed video format, the system stops the running tasks associated with the video and returns the proper error code to the client.

Typical errors for each system component are covered by the following playbook:

- Upload error: retry a few times.
- Split video error: if older versions of clients cannot split videos by GOP alignment, the entire video is passed to the server. The job of splitting videos is done on the server-side.
- Transcoding error: retry.
- Preprocessor error: regenerate DAG diagram.
- DAG scheduler error: reschedule a task.
- Resource manager queue down: use a replica.
- Task worker down: retry the task on a new worker.
- API server down: API servers are stateless so requests will be directed to a different API server.
- Metadata cache server down: data is replicated multiple times. If one node goes down, you can still access other nodes to fetch data. We can bring up a new cache server to replace the dead one.
- Metadata DB server down:
  o Master is down. If the master is down, promote one of the slaves to act as the new master.

o   Slave is down. If a slave goes down, you can use another slave for reads and bring up another database server to replace the dead one.

Step 4 - Wrap up

In this chapter, we presented the architecture design for video streaming services like YouTube. If there is extra time at the end of the interview, here are a few additional points:

- Scale the API tier: Because API servers are stateless, it is easy to scale API tier horizontally.

- Scale the database: You can talk about database replication and sharding.

- Live streaming: It refers to the process of how a video is recorded and broadcasted in real time. Although our system is not designed specifically for live streaming, live streaming and non-live streaming have some similarities: both require uploading, encoding, and streaming. The notable differences are:

  o   Live streaming has a higher latency requirement, so it might need a different streaming protocol.

  o   Live streaming has a lower requirement for parallelism because small chunks of data are already processed in real-time.

  o   Live streaming requires different sets of error handling. Any error handling that takes too much time is not acceptable.

- Video takedowns: Videos that violate copyrights, pornography, or other illegal acts shall be removed. Some can be discovered by the system during the upload process, while others might be discovered through user flagging.

Congratulations on getting this far! Now give yourself a pat on the back. Good job!

Reference materials

[1] YouTube by the numbers: https://www.omnicoreagency.com/
youtube-statistics/

[2] 2019 YouTube Demographics:
https://blog.hubspot.com/marketing/youtube-demographics

[3] Cloudfront Pricing: https://aws.amazon.com/cloudfront/pricing/

[4] Netflix on AWS:
https://aws.amazon.com/solutions/case-studies/netflix/

[5] Akamai homepage: https://www.akamai.com/

[6] Binary large object:
https://en.wikipedia.org/wiki/Binary_large_object

[7] Here's What You Need to Know About Streaming Protocols:
https://www.dacast.com/blog/streaming-protocols/

[8] SVE: Distributed Video Processing at Facebook Scale:
https://www.cs.princeton.edu/~wlloyd/papers/sve-sosp17.pdf

[9] Weibo video processing architecture (in Chinese):
https://www.upyun.com/opentalk/399.html

[10] Delegate access with a shared access signature:
https://docs.microsoft.com/en-us/rest/api/storageservices/delegate-
access-with-shared-access-signature

[11] YouTube scalability talk by early YouTube employee:
https://www.youtube.com/watch?v=w5WVu624fY8

[12] Understanding the characteristics of internet short video sharing: A
youtube-based measurement study. https://arxiv.org/pdf/0707.3670.pdf

[13] Content Popularity for Open Connect:
https://netflixtechblog.com/content-popularity-for-open-connect-
b86d56f613b

15

DESIGN GOOGLE DRIVE

In recent years, cloud storage services such as Google Drive, Dropbox, Microsoft OneDrive, and Apple iCloud have become very popular. In this chapter, you are asked to design Google Drive.

Let us take a moment to understand Google Drive before jumping into the design. Google Drive is a file storage and synchronization service that helps you store documents, photos, videos, and other files in the cloud. You can access your files from any computer, smartphone, and tablet. You can easily share those files with friends, family, and coworkers [1]. Figure 15-1 and 15-2 show what Google drive looks like on a browser and mobile application, respectively.

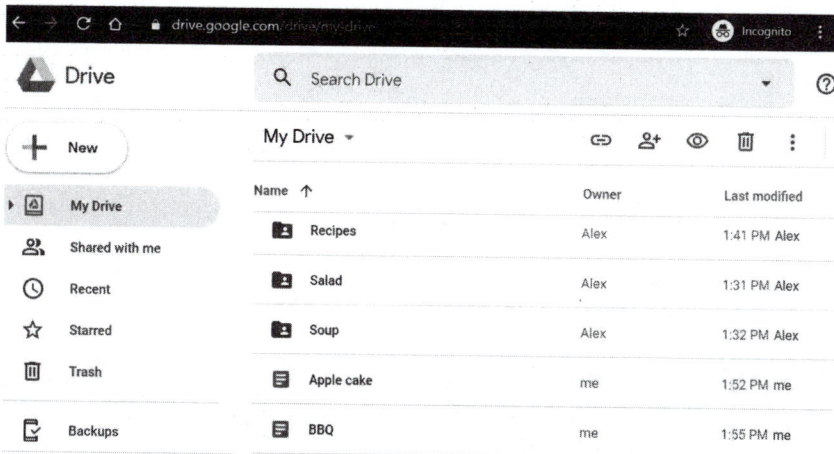

Figure 15-1

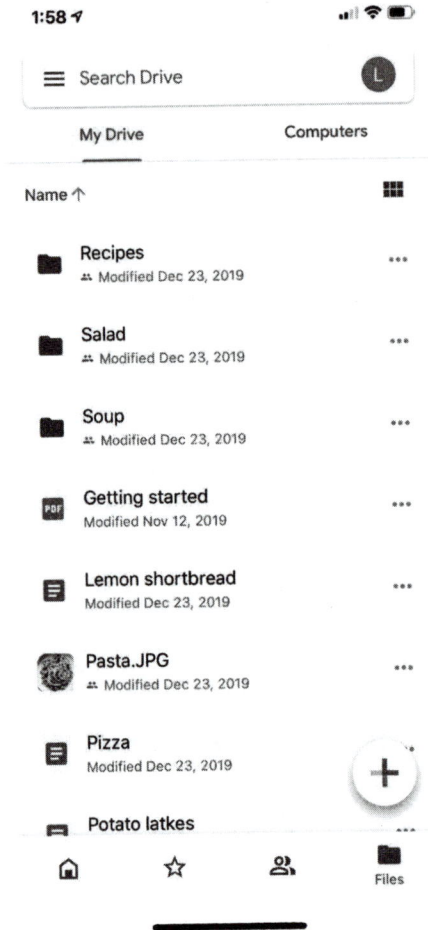

Figure 15-2

Step 1 - Understand the problem and establish design scope

Designing a Google drive is a big project, so it is important to ask questions to narrow down the scope.

Candidate: What are the most important features?
Interviewer: Upload and download files, file sync, and notifications.

Candidate: Is this a mobile app, a web app, or both?
Interviewer: Both.

Candidate: What are the supported file formats?
Interviewer: Any file type.

Candidate: Do files need to be encrypted?
Interview: Yes, files in the storage must be encrypted.

Candidate: Is there a file size limit?
Interview: Yes, files must be 10 GB or smaller.

Candidate: How many users does the product have?
Interviewer: 10M DAU.

In this chapter, we focus on the following features:

- Add files. The easiest way to add a file is to drag and drop a file into Google drive.

- Download files.

- Sync files across multiple devices. When a file is added to one device, it is automatically synced to other devices.

- See file revisions.

- Share files with your friends, family, and coworkers

- Send a notification when a file is edited, deleted, or shared with you.

Features not discussed in this chapter include:

- Google doc editing and collaboration. Google doc allows multiple people to edit the same document simultaneously. This is out of our design scope.

Other than clarifying requirements, it is important to understand non-functional requirements:

- Reliability. Reliability is extremely important for a storage system. Data loss is unacceptable.

- Fast sync speed. If file sync takes too much time, users will become impatient and abandon the product.

- Bandwidth usage. If a product takes a lot of unnecessary network bandwidth, users will be unhappy, especially when they are on a mobile data plan.

- Scalability. The system should be able to handle high volumes of traffic.

- High availability. Users should still be able to use the system when some servers are offline, slowed down, or have unexpected network errors.

Back of the envelope estimation

- Assume the application has 50 million signed up users and 10 million DAU.

- Users get 10 GB free space.

- Assume users upload 2 files per day. The average file size is 500 KB.

- 1:1 read to write ratio.

- Total space allocated: 50 million * 10 GB = 500 Petabyte

- QPS for upload API: 10 million * 2 uploads / 24 hours / 3600 seconds = ~ 240

- Peak QPS = QPS * 2 = 480

Step 2 - Propose high-level design and get buy-in

Instead of showing the high-level design diagram from the beginning, we will use a slightly different approach. We will start with something simple: build everything in a single server. Then, gradually scale it up to support millions of users. By doing this exercise, it will refresh your memory about some important topics covered in the book.

Let us start with a single server setup as listed below:

- A web server to upload and download files.

- A database to keep track of metadata like user data, login info, files info, etc.

- A storage system to store files. We allocate 1TB of storage space to store files.

We spend a few hours setting up an Apache web server, a MySql database, and a directory called *drive/* as the root directory to store uploaded files. Under *drive/* directory, there is a list of directories, known as namespaces. Each namespace contains all the uploaded files for that user. The filename on the server is kept the same as the original file name. Each file or folder can be uniquely identified by joining the namespace and the relative path.

Figure 15-3 shows an example of how the */drive* directory looks like on the left side and its expanded view on the right side.

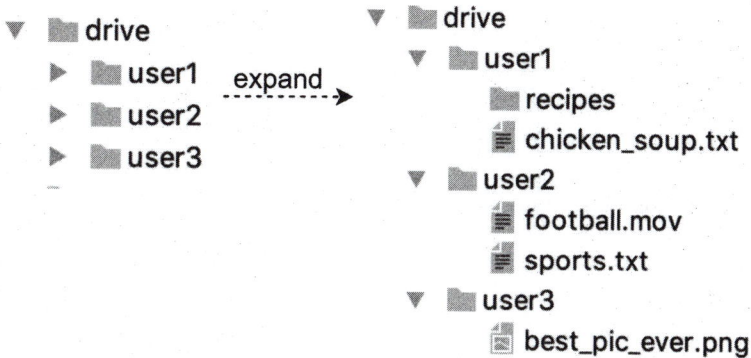

Figure 15-3

APIs

What do the APIs look like? We primary need 3 APIs: upload a file, download a file, and get file revisions.

1. Upload a file to Google Drive

Two types of uploads are supported:

- Simple upload. Use this upload type when the file size is small.
- Resumable upload. Use this upload type when the file size is large and there is high chance of network interruption.

Here is an example of resumable upload API:

https://api.example.com/files/upload?uploadType=resumable

Params:

- uploadType=resumable
- data: Local file to be uploaded.

A resumable upload is achieved by the following 3 steps [2]:

- Send the initial request to retrieve the resumable URL.
- Upload the data and monitor upload state.
- If upload is disturbed, resume the upload.

2. Download a file from Google Drive

Example API: https://api.example.com/files/download

Params:

- path: download file path.

  Example params:
  {

  "path": "/recipes/soup/best_soup.txt"

  }

3. Get file revisions

Example API: https://api.example.com/files/list_revisions

Params:

- path: The path to the file you want to get the revision history.
- limit: The maximum number of revisions to return.

  Example params:

  {

  "path": "/recipes/soup/best_soup.txt",

  "limit": 20

  }

All the APIs require user authentication and use HTTPS. Secure Sockets Layer (SSL) protects data transfer between the client and backend servers.

Move away from single server

As more files are uploaded, eventually you get the space full alert as shown in Figure 15-4.

Figure 15-4

Only 10 MB of storage space is left! This is an emergency as users cannot upload files anymore. The first solution comes to mind is to shard the data, so it is stored on multiple storage servers. Figure 15-5 shows an example of sharding based on *user_id*.

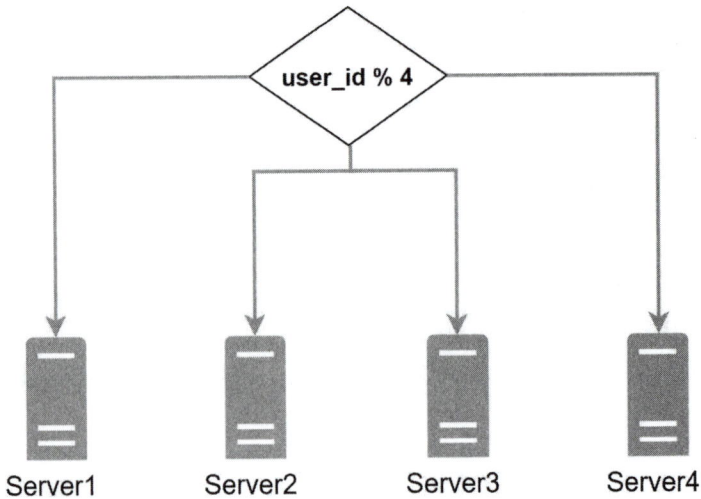

Figure 15-5

You pull an all-nighter to set up database sharding and monitor it closely. Everything works smoothly again. You have stopped the fire, but you are still worried about potential data losses in case of storage server outage. You ask around and your backend guru friend Frank told you that many leading companies like Netflix and Airbnb use Amazon S3 for storage. "Amazon Simple Storage Service (Amazon S3) is an object storage service that offers industry-leading scalability, data availability, security, and performance" [3]. You decide to do some research to see if it is a good fit.

After a lot of reading, you gain a good understanding of the S3 storage system and decide to store files in S3. Amazon S3 supports same-region and cross-region replication. A region is a geographic area where Amazon web services (AWS) have data centers. As shown in Figure 15-6, data can be replicated on the same-region (left side) and cross-region (right side). Redundant files are stored in multiple regions to guard against data loss and ensure availability. A bucket is like a folder in file systems.

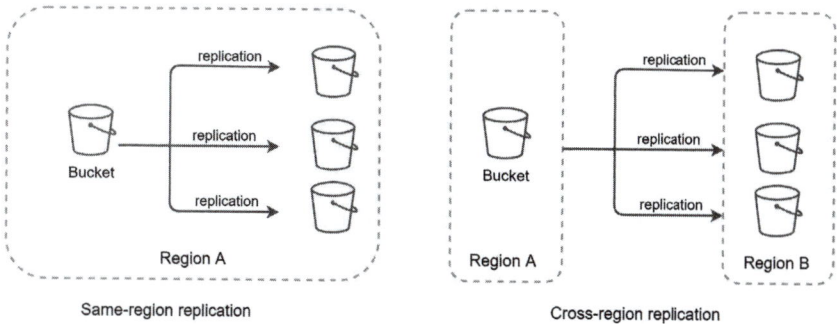

Figure 15-6

After putting files in S3, you can finally have a good night's sleep without worrying about data losses. To stop similar problems from happening in the future, you decide to do further research on areas you can improve. Here are a few areas you find:

- Load balancer: Add a load balancer to distribute network traffic. A load balancer ensures evenly distributed traffic, and if a web server goes down, it will redistribute the traffic.

- Web servers: After a load balancer is added, more web servers can be added/removed easily, depending on the traffic load.

- Metadata database: Move the database out of the server to avoid single point of failure. In the meantime, set up data replication and sharding to meet the availability and scalability requirements.

- File storage: Amazon S3 is used for file storage. To ensure availability and durability, files are replicated in two separate geographical regions.

After applying the above improvements, you have successfully decoupled web servers, metadata database, and file storage from a single server. The updated design is shown in Figure 15-7.

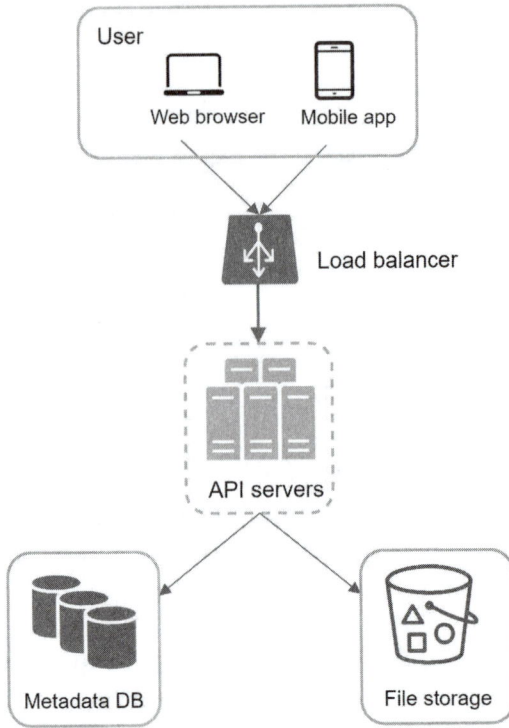

Figure 15-7

Sync conflicts

For a large storage system like Google Drive, sync conflicts happen from time to time. When two users modify the same file or folder at the same time, a conflict happens. How can we resolve the conflict? Here is our strategy: the first version that gets processed wins, and the version that gets processed later receives a conflict. Figure 15-8 shows an example of a sync conflict.

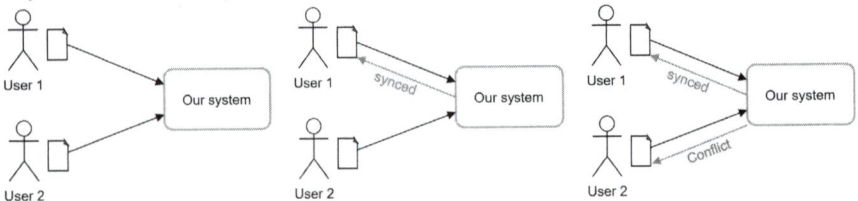

Figure 15-8

In Figure 15-8, user 1 and user 2 tries to update the same file at the same time, but user 1's file is processed by our system first. User 1's update operation goes through, but, user 2 gets a sync conflict. How can we resolve the conflict for user 2? Our system presents both copies of the same file: user 2's local copy and the latest version from the server (Figure 15-9). User 2 has the option to merge both files or override one version with the other.

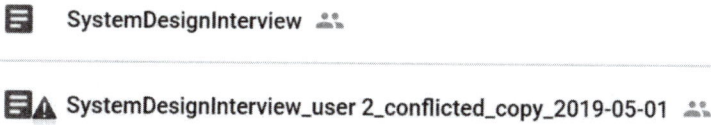

SystemDesignInterview

SystemDesignInterview_user 2_conflicted_copy_2019-05-01

Figure 15-9

While multiple users are editing the same document at the same, it is challenging to keep the document synchronized. Interested readers should refer to the reference material [4] [5].

High-level design

Figure 15-10 illustrates the proposed high-level design. Let us examine each component of the system.

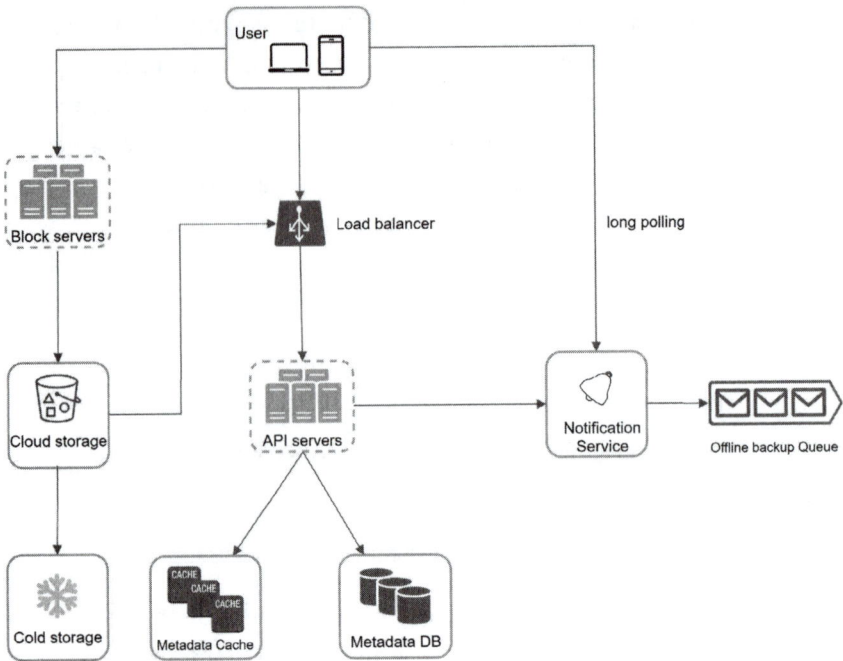

Figure 15-10

User: A user uses the application either through a browser or mobile app.

Block servers: Block servers upload blocks to cloud storage. Block storage, referred to as block-level storage, is a technology to store data files on cloud-based environments. A file can be split into several blocks, each with a unique hash value, stored in our metadata database. Each block is treated as an independent object and stored in our storage system (S3). To reconstruct a file, blocks are joined in a particular order. As for the block size, we use Dropbox as a reference: it sets the maximal size of a block to 4MB [6].

Cloud storage: A file is split into smaller blocks and stored in cloud storage.

Cold storage: Cold storage is a computer system designed for storing inactive data, meaning files are not accessed for a long time.

Load balancer: A load balancer evenly distributes requests among API servers.

API servers: These are responsible for almost everything other than the uploading flow. API servers are used for user authentication, managing user profile, updating file metadata, etc.

Metadata database: It stores metadata of users, files, blocks, versions, etc. Please note that files are stored in the cloud and the metadata database only contains metadata.

Metadata cache: Some of the metadata are cached for fast retrieval.

Notification service: It is a publisher/subscriber system that allows data to be transferred from notification service to clients as certain events happen. In our specific case, notification service notifies relevant clients when a file is added/edited/removed elsewhere so they can pull the latest changes.

Offline backup queue: If a client is offline and cannot pull the latest file changes, the offline backup queue stores the info so changes will be synced when the client is online.

We have discussed the design of Google Drive at the high-level. Some of the components are complicated and worth careful examination; we will discuss these in detail in the deep dive.

Step 3 - Design deep dive

In this section, we will take a close look at the following: block servers, metadata database, upload flow, download flow, notification service, save storage space and failure handling.

Block servers

For large files that are updated regularly, sending the whole file on each update consumes a lot of bandwidth. Two optimizations are proposed to minimize the amount of network traffic being transmitted:

- Delta sync. When a file is modified, only modified blocks are synced instead of the whole file using a sync algorithm [7] [8].

- Compression. Applying compression on blocks can significantly reduce the data size. Thus, blocks are compressed using compression algorithms depending on file types. For example, gzip and bzip2 are used to compress text files. Different compression algorithms are needed to compress images and videos.

In our system, block servers do the heavy lifting work for uploading files. Block servers process files passed from clients by splitting a file into blocks, compressing each block, and encrypting them. Instead of uploading the whole file to the storage system, only modified blocks are transferred.

Figure 15-11 shows how a block server works when a new file is added.

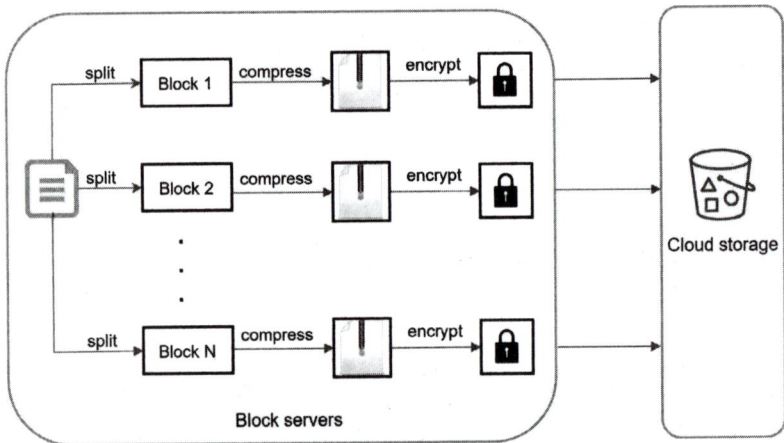

Figure 15-11

- A file is split into smaller blocks.
- Each block is compressed using compression algorithms.
- To ensure security, each block is encrypted before it is sent to cloud storage.
- Blocks are uploaded to the cloud storage.

Figure 15-12 illustrates delta sync, meaning only modified blocks are transferred to cloud storage. Highlighted blocks "block 2" and "block

5" represent changed blocks. Using delta sync, only those two blocks are uploaded to the cloud storage.

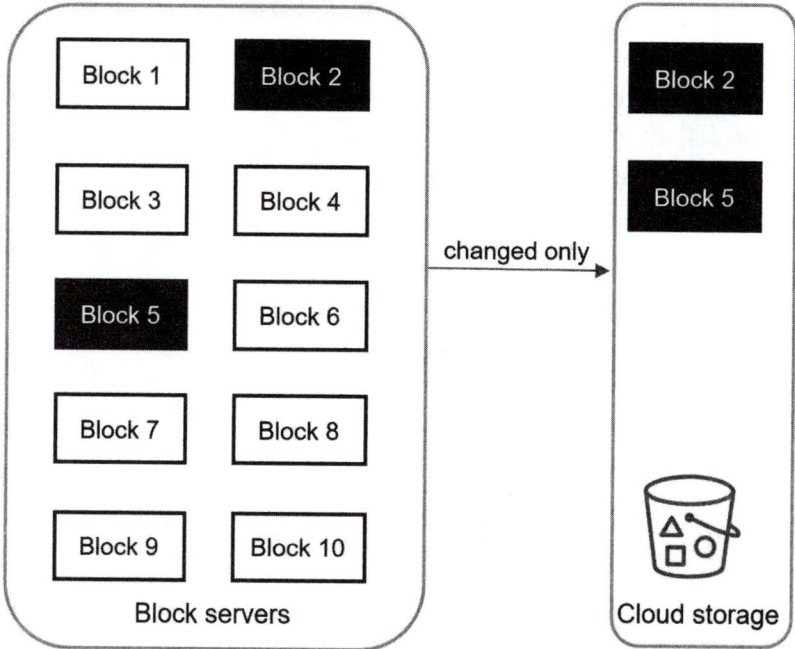

Figure 15-12

Block servers allow us to save network traffic by providing delta sync and compression.

High consistency requirement

Our system requires strong consistency by default. It is unacceptable for a file to be shown differently by different clients at the same time. The system needs to provide strong consistency for metadata cache and database layers.

Memory caches adopt an eventual consistency model by default, which means different replicas might have different data. To achieve strong consistency, we must ensure the following:

- Data in cache replicas and the master is consistent.

- Invalidate caches on database write to ensure cache and database hold the same value.

Achieving strong consistency in a relational database is easy because it maintains the ACID (Atomicity, Consistency, Isolation, Durability) properties [9]. However, NoSQL databases do not support ACID properties by default. ACID properties must be programmatically incorporated in synchronization logic. In our design, we choose relational databases because the ACID is natively supported.

Metadata database

Figure 15-13 shows the database schema design. Please note this is a highly simplified version as it only includes the most important tables and interesting fields.

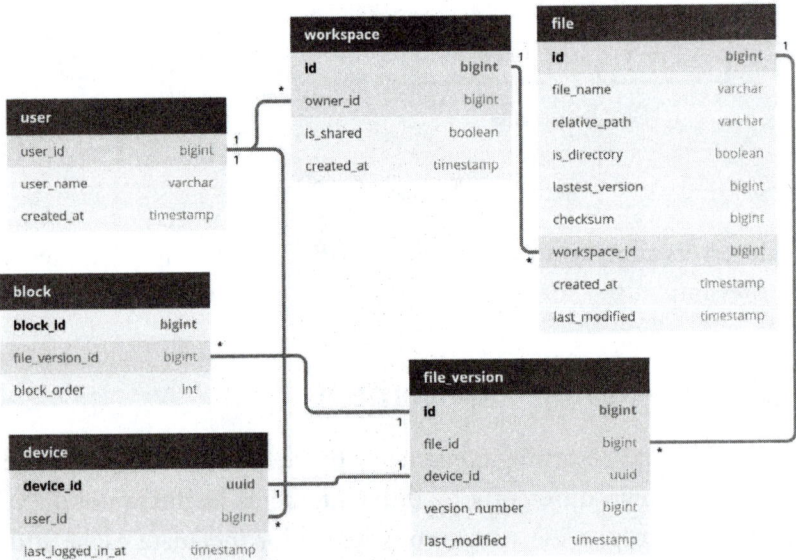

Figure 15-13

User: The user table contains basic information about the user such as username, email, profile photo, etc.

Device: Device table stores device info. *Push_id* is used for sending and receiving mobile push notifications. Please note a user can have multiple devices.

Namespace: A namespace is the root directory of a user.

File: File table stores everything related to the latest file.

File_version: It stores version history of a file. Existing rows are read-only to keep the integrity of the file revision history.

Block: It stores everything related to a file block. A file of any version can be reconstructed by joining all the blocks in the correct order.

Upload flow

Let us discuss what happens when a client uploads a file. To better understand the flow, we draw the sequence diagram as shown in Figure 15-14.

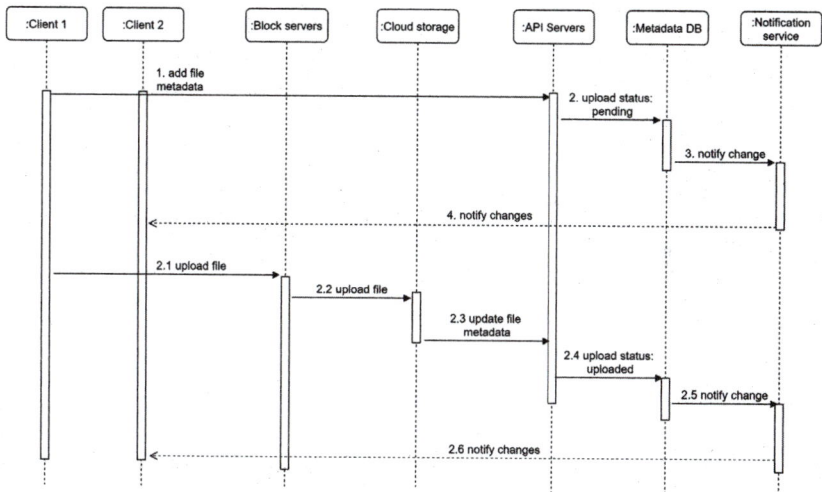

Figure 15-14

In Figure 15-14, two requests are sent in parallel: add file metadata and upload the file to cloud storage. Both requests originate from client 1.

- Add file metadata.
  1. Client 1 sends a request to add the metadata of the new file.

2. Store the new file metadata in metadata DB and change the file upload status to "pending."

3. Notify the notification service that a new file is being added.

4. The notification service notifies relevant clients (client 2) that a file is being uploaded.

- Upload files to cloud storage.

2.1 Client 1 uploads the content of the file to block servers.

2.2 Block servers chunk the files into blocks, compress, encrypt the blocks, and upload them to cloud storage.

2.3 Once the file is uploaded, cloud storage triggers upload completion callback. The request is sent to API servers.

2.4 File status changed to "uploaded" in Metadata DB.

2.5 Notify the notification service that a file status is changed to "uploaded."

2.6 The notification service notifies relevant clients (client 2) that a file is fully uploaded.

When a file is edited, the flow is similar, so we will not repeat it.

Download flow

Download flow is triggered when a file is added or edited elsewhere. How does a client know if a file is added or edited by another client? There are two ways a client can know:

- If client A is online while a file is changed by another client, notification service will inform client A that changes are made somewhere so it needs to pull the latest data.

- If client A is offline while a file is changed by another client, data will be saved to the cache. When the offline client is online again, it pulls the latest changes.

Once a client knows a file is changed, it first requests metadata via API servers, then downloads blocks to construct the file. Figure 15-15 shows the detailed flow. Note, only the most important components are shown in the diagram due to space constraint.

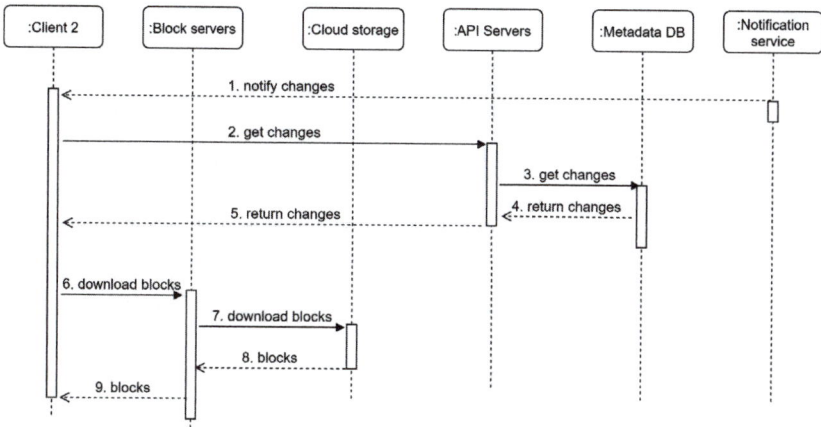

Figure 15-15

1. Notification service informs client 2 that a file is changed somewhere else.

2. Once client 2 knows that new updates are available, it sends a request to fetch metadata.

3. API servers call metadata DB to fetch metadata of the changes.

4. Metadata is returned to the API servers.

5. Client 2 gets the metadata.

6. Once the client receives the metadata, it sends requests to block servers to download blocks.

7. Block servers first download blocks from cloud storage.

8. Cloud storage returns blocks to the block servers.

9. Client 2 downloads all the new blocks to reconstruct the file.

Notification service

To maintain file consistency, any mutation of a file performed locally needs to be informed to other clients to reduce conflicts. Notification service is built to serve this purpose. At the high-level, notification service allows data to be transferred to clients as events happen. Here are a few options:

- Long polling. Dropbox uses long polling [10].

- WebSocket. WebSocket provides a persistent connection between the client and the server. Communication is bi-directional.

Even though both options work well, we opt for long polling for the following two reasons:

- Communication for notification service is not bi-directional. The server sends information about file changes to the client, but not vice versa.

- WebSocket is suited for real-time bi-directional communication such as a chat app. For Google Drive, notifications are sent infrequently with no burst of data.

With long polling, each client establishes a long poll connection to the notification service. If changes to a file are detected, the client will close the long poll connection. Closing the connection means a client must connect to the metadata server to download the latest changes. After a response is received or connection timeout is reached, a client immediately sends a new request to keep the connection open.

Save storage space

To support file version history and ensure reliability, multiple versions of the same file are stored across multiple data centers. Storage space can be filled up quickly with frequent backups of all file revisions. Three techniques are proposed to reduce storage costs:

- De-duplicate data blocks. Eliminating redundant blocks at the account level is an easy way to save space. Two blocks are identical if they have the same hash value.

- Adopt an intelligent data backup strategy. Two optimization strategies can be applied:

  o Set a limit: We can set a limit for the number of versions to store. If the limit is reached, the oldest version will be replaced with the new version.

o   Keep valuable versions only: Some files might be edited frequently. For example, saving every edited version for a heavily modified document could mean the file is saved over 1000 times within a short period. To avoid unnecessary copies, we could limit the number of saved versions. We give more weight to recent versions. Experimentation is helpful to figure out the optimal number of versions to save.

- Moving infrequently used data to cold storage. Cold data is the data that has not been active for months or years. Cold storage like Amazon S3 glacier [11] is much cheaper than S3.

Failure Handling

Failures can occur in a large-scale system and we must adopt design strategies to address these failures. Your interviewer might be interested in hearing about how you handle the following system failures:

- Load balancer failure: If a load balancer fails, the secondary would become active and pick up the traffic. Load balancers usually monitor each other using a heartbeat, a periodic signal sent between load balancers. A load balancer is considered as failed if it has not sent a heartbeat for some time.

- Block server failure: If a block server fails, other servers pick up unfinished or pending jobs.

- Cloud storage failure: S3 buckets are replicated multiple times in different regions. If files are not available in one region, they can be fetched from different regions.

- API server failure: It is a stateless service. If an API server fails, the traffic is redirected to other API servers by a load balancer.

- Metadata cache failure: Metadata cache servers are replicated multiple times. If one node goes down, you can still access other nodes to fetch data. We will bring up a new cache server to replace the failed one.

- Metadata DB failure.

  o   Master down: If the master is down, promote one of the slaves to act as a new master and bring up a new slave node.

  o   Slave down: If a slave is down, you can use another slave for read operations and bring another database server to replace the failed one.

- Notification service failure: Every online user keeps a long poll connection with the notification server. Thus, each notification server is connected with many users. According to the Dropbox talk in 2012 [6], over 1 million connections are open per machine. If a server goes down, all the long poll connections are lost so clients must reconnect to a different server. Even though one server can keep many open connections, it cannot reconnect all the lost connections at once. Reconnecting with all the lost clients is a relatively slow process.

- Offline backup queue failure: Queues are replicated multiple times. If one queue fails, consumers of the queue may need to re-subscribe to the backup queue.

Step 4 - Wrap up

In this chapter, we proposed a system design to support Google Drive. The combination of strong consistency, low network bandwidth and fast sync make the design interesting. Our design contains two flows: manage file metadata and file sync. Notification service is another important component of the system. It uses long polling to keep clients up to date with file changes.

Like any system design interview questions, there is no perfect solution. Every company has its unique constraints and you must design a system to fit those constraints. Knowing the tradeoffs of your design and technology choices are important. If there are a few minutes left, you can talk about different design choices.

For example, we can upload files directly to cloud storage from the client instead of going through block servers. The advantage of this approach is that it makes file upload faster because a file only needs to be transferred once to the cloud storage. In our design, a file is transferred to block servers first, and then to the cloud storage. However, the new approach has a few drawbacks:

- First, the same chunking, compression, and encryption logic must be implemented on different platforms (iOS, Android, Web). It is error-prone and requires a lot of engineering effort. In our design, all those logics are implemented in a centralized place: block servers.

- Second, as a client can easily be hacked or manipulated, implementing encrypting logic on the client side is not ideal.

Another interesting evolution of the system is moving online/offline logic to a separate service. Let us call it presence service. By moving presence service out of notification servers, online/offline functionality can easily be integrated by other services.

Congratulations on getting this far! Now give yourself a pat on the back. Good job!

Reference materials

[1] Google Drive: https://www.google.com/drive/

[2] Upload file data:
https://developers.google.com/drive/api/v2/manage-uploads

[3] Amazon S3: https://aws.amazon.com/s3

[4] Differential Synchronization https://neil.fraser.name/writing/sync/

[5] Differential Synchronization YouTube talk:
https://www.youtube.com/watch?v=S2Hp_1jqpY8

[6] How We've Scaled Dropbox: https://youtu.be/PE4gwstWhmc

[7] Tridgell, A., & Mackerras, P. (1996). The rsync algorithm.

[8] Librsync. (n.d.). Retrieved April 18, 2015, from
https://github.com/librsync/librsync

[9] ACID: https://en.wikipedia.org/wiki/ACID

[10] Dropbox security white paper:
https://www.dropbox.com/static/business/resources/Security_
Whitepaper.pdf

[11] Amazon S3 Glacier: https://aws.amazon.com/glacier/faqs/

16

THE LEARNING CONTINUES

Designing good systems requires years of accumulation of knowledge. One shortcut is to dive into real-world system architectures. Below is a collection of helpful reading materials. We highly recommend you pay attention to both the shared principles and the underlying technologies. Researching each technology and understanding what problems it solves is a great way to strengthen your knowledge base and refine the design process.

Real-world systems

The following materials can help you understand general design ideas of real system architectures behind different companies.

Facebook Timeline: Brought To You By The Power Of Denormalization: https://goo.gl/FCNrbm

Scale at Facebook: https://goo.gl/NGTdCs

Building Timeline: Scaling up to hold your life story: https://goo.gl/8p5wDV

Erlang at Facebook (Facebook chat): https://goo.gl/zSLHrj

Facebook Chat: https://goo.gl/qzSiWC

Finding a needle in Haystack: Facebook's photo storage: https://goo.gl/edj4FL

Serving Facebook Multifeed: Efficiency, performance gains through re-design: https://goo.gl/adFVMQ

Scaling Memcache at Facebook: https://goo.gl/rZiAhX

TAO: Facebook's Distributed Data Store for the Social Graph:
https://goo.gl/Tk1DyH

Amazon Architecture: https://goo.gl/k4feoW

Dynamo: Amazon's Highly Available Key-value Store:
https://goo.gl/C7zxDL

A 360 Degree View Of The Entire Netflix Stack: https://goo.gl/rYSDTz

It's All A/Bout Testing: The Netflix Experimentation Platform:
https://goo.gl/agbA4K

Netflix Recommendations: Beyond the 5 stars (Part 1):
https://goo.gl/A4FkYi

Netflix Recommendations: Beyond the 5 stars (Part 2):
https://goo.gl/XNPMXm

Google Architecture: https://goo.gl/dvkDiY

The Google File System (Google Docs): https://goo.gl/xj5n9R

Differential Synchronization (Google Docs): https://goo.gl/9zqG7x

YouTube Architecture: https://goo.gl/mCPRUF

Seattle Conference on Scalability: YouTube Scalability:
https://goo.gl/dH3zYq

Bigtable: A Distributed Storage System for Structured Data:
https://goo.gl/6NaZca

Instagram Architecture: 14 Million Users, Terabytes Of Photos, 100s
Of Instances, Dozens Of Technologies: https://goo.gl/s1VcW5

The Architecture Twitter Uses To Deal With 150M Active Users:
https://goo.gl/EwvfRd

Scaling Twitter: Making Twitter 10000 Percent Faster:
https://goo.gl/nYGC1k

Announcing Snowflake (Snowflake is a network service for generating
unique ID numbers at high scale with some simple guarantees):
https://goo.gl/GzVWYm

Timelines at Scale: https://goo.gl/8KbqTy

How Uber Scales Their Real-Time Market Platform:
https://goo.gl/kGZuVy

Scaling Pinterest: https://goo.gl/KtmjW3

Pinterest Architecture Update: https://goo.gl/w6rRsf

A Brief History of Scaling LinkedIn: https://goo.gl/8A1Pi8

Flickr Architecture: https://goo.gl/dWtgYa

How We've Scaled Dropbox: https://goo.gl/NjBDtC

The WhatsApp Architecture Facebook Bought For $19 Billion:
https://bit.ly/2AHJnFn

Company engineering blogs

If you are going to interview with a company, it is a great idea to read
their engineering blogs and get familiar with technologies and systems
adopted and implemented there. Besides, engineering blogs provide in-
valuable insights about certain fields. Reading them regularly could help
us become better engineers.

Here is a list of engineering blogs of well-known large companies
and startups.

Airbnb: https://medium.com/airbnb-engineering

Amazon: https://developer.amazon.com/blogs

Asana: https://blog.asana.com/category/eng

Atlassian: https://developer.atlassian.com/blog

Bittorrent: http://engineering.bittorrent.com

Cloudera: https://blog.cloudera.com

Docker: https://blog.docker.com

Dropbox: https://blogs.dropbox.com/tech

eBay: http://www.ebaytechblog.com

Facebook: https://code.facebook.com/posts

GitHub: https://githubengineering.com

Google: https://developers.googleblog.com

Groupon: https://engineering.groupon.com

Highscalability: http://highscalability.com

Instacart: https://tech.instacart.com

Instagram: https://engineering.instagram.com

Linkedin: https://engineering.linkedin.com/blog

Mixpanel: https://mixpanel.com/blog

Netflix: https://medium.com/netflix-techblog

Nextdoor: https://engblog.nextdoor.com

PayPal: https://www.paypal-engineering.com

Pinterest: https://engineering.pinterest.com

Quora: https://engineering.quora.com

Reddit: https://redditblog.com

Salesforce: https://developer.salesforce.com/blogs/engineering

Shopify: https://engineering.shopify.com

Slack: https://slack.engineering

Soundcloud: https://developers.soundcloud.com/blog

Spotify: https://labs.spotify.com

Square: https://stripe.com/blog/engineering

Stripe: https://developer.squareup.com/blog/

System design primer:
https://github.com/donnemartin/system-design-primer

Twitter: https://blog.twitter.com/engineering/en_us.html

Thumbtack: https://www.thumbtack.com/engineering

Uber: http://eng.uber.com

Yahoo: https://yahooeng.tumblr.com

Yelp: https://engineeringblog.yelp.com

Zoom: https://medium.com/zoom-developer-blog

AFTERWORD

Congratulations! You are at the end of this interview guide. You have accumulated skills and knowledge to design systems. Not everyone has the discipline to learn what you have learned. Take a moment and pat yourself on the back. Your hard work will be paid off.

Landing a dream job is a long journey and requires lots of time and effort. Practice makes perfect. Best luck!

Thank you for buying and reading this book. Without readers like you, our work would not exist. We hope you have enjoyed the book!

If you don't mind, please review this book on Amazon: http://bit.ly/sysreview8 It would help me attract more wonderful readers like you.

System Design Newsletter

Subscribe to the ByteByteGo weekly newsletter to get a Free System Design PDF (158 pages): blog.bytebytego.com

EP26: Proxy vs reverse proxy

In this issue, we will cover: Why is Nginx called a "reverse" proxy? CAP theorem How Does Live Streaming Platform Work? CDN Postman the API platform for...

ALEX XU   OCT 1   ♡ 225   💬 6   ↪

EP17: Design patterns cheat sheet. Also...

For this week's newsletter, we will cover: Design patterns cheat sheet 6 ways to turn code into beautiful architecture diagrams What is a File...

ALEX XU   JUL 30   ♡ 166   💬 7   ↪

EP22: Latency numbers you should know. Also...

In this newsletter, we'll cover the following topics: Latency numbers you should know Microservice architecture Handling hotspot accounts E-commerce...

ALEX XU   SEP 3   ♡ 153   💬 9   ↪

EP15: What happens when you swipe a credit card? Also...

For this week's newsletter, we will cover: How does VISA work when we swipe a credit card at a merchant's shop? What are the differences between bare...

ALEX XU   JUL 16   ♡ 141   💬 8   ↪

EP14: Algorithms you should know for System Design. Also...

In this newsletter, we'll cover the following topics: Algorithms you should know before taking System Design Interviews How to store passwords safely in...

ALEX XU   JUL 9   ♡ 185   💬 2   ↪